Physics

Arthur Gibbons

Educational Consultant

Published by BBC Educational Publishing,
BBC White City, 201 Wood Lane, London W12 7TS
First published 2000, Reprinted 2001, Reprinted 2002

ISBN: 0 563 47492 0

Designed by Linda Reed and Associates
Illustrations by Tech Type
Examination questions reproduced with the permission of SQA
Printed in Great Britain by Bell & Bain Ltd., Glasgow

BBC

Contents

Introduction

About BITESIZE Physics

STOP! Don't skip over these introductory pages – they tell you about the purpose of this book and how you should use it.

The book has been written to help you **revise** and **prepare** for your Standard Grade Physics exam. In it you will find summaries of the essential facts, formulae, rules and relationships that you need to know for the examination.

BITESIZE Physics is a revision guide, a work book, a note book and a memo pad. Use it in whatever way helps you with your revision. The book is not a text book. It assumes that you have already studied the Standard Grade Physics course.

The book is designed to put the icing on the cake. It will enable you to brush up on the **knowledge** you need for the exam. It will also help to improve your **understanding** of the course content. Good knowledge and understanding are essential for **solving problems**. For example, if you have little knowledge of the physics quantities current, voltage and resistance and no real understanding of what these terms mean, then you are unlikely to be able to solve problems on electricity.

The book is called 'bitesize' because the best way to revise is in bitesized chunks, not all in one go. It contains 30 bitesized sections. Each section – on a double page – will take 30–40 minutes to work through. That adds up to about 15–20 hours of revision time in all. Spread this over the eight weeks leading up to your exam and you are looking at about two hours of Physics revision per week. Not a lot to ask – and still plenty of time for your other subjects and interests.

About the book

The book is divided into six sections:

- Section 1: Forces and motion
- Section 2: Energy
- Section 3: Electricity
- Section 4: Electronics
- Section 5: Waves
- Section 6: Radioactivity

The sections cover most of the topics you should revise for both the General Level and the Credit Level exams.

For the Credit Level exam, you are expected to know and understand more than at General Level. The Credit Level exam includes the content for General Level. So if you are revising for Credit Level, **all** of the revision material is for you.

KEY TO SYMBOLS

- Ⓒ Credit Level only
- A question to think about
- An activity to do
- A link to the video

If you are tackling only the General Level exam, you might want to skip some parts of a section. It is up to you.

Each section begins with an introduction and a FactZone. The introduction tells you what you should be able to do when you have completed the section. It also includes a summary of the section's contents. In the summary the key physics words are in bold. Make sure you know what these words mean – ask your teacher if you are not sure.

The FactZone page gives a useful summary of all the facts and formulae that are covered by a section. Use it as a checklist to make sure that you have understood all of the information given in the section. Discuss the activities with a friend or your teacher if you are not sure of what you have to do.

Each section invites you to participate actively in your revision. You are given **activities to do, questions to think about** and **practice questions**. The answers to the questions to think about and the practice questions are given at the end of the book (pages 92-93), so that you can check on how well your revision is going. Answering these questions will help you to practise the skills you need for the exam and will make you to concentrate in order to get the maximum benefit from your revision.

At the end of each section there are also some questions from past Standard Grade exam papers in Physics. Once you have completed a whole section, try these exam questions. You can practise exam technique by working under exam conditions. (See pages 6 and 7 for tips on exam technique.)

Once you have completed the questions, mark your answers using the marking guide given (see pages 94 to 96.) Be honest with yourself! Use the results of your self-assessment to decide if you need to revise a section or a bitesized chunk further. You might want to look back over your notes or in your text book, or perhaps ask your teacher for some help.

The REMEMBER! paragraphs mention important points that you should note.

Television and online revision

Although the book can be used separately as a revision aid, it has been designed to link with a television programme produced by the BBC. The sections covered by the book are all dealt with in the BBC Bitesize Standard Grade Physics TV programmes. Like this book, the programme is in bitesized chunks.

You can record the TV programme when it is broadcast on BBC2. Then, while you are watching the recording of the programme, you can note down the time code from the video at the relevant place in the book, so you can quickly find that section again.

The book also links with the BBC Bitesize website. On the site, you will find both revision and test material for Standard Grade Physics along with tips and an 'Ask a teacher' facility.

THE ONLINE SERVICE

You can find extra support, tips and answers to your exam queries on the Standard Grade BITESIZE website. The address is www.bbc.co.uk/scotland/revision/physics

REMEMBER
The purpose of the exam is to find out what knowledge of Physics you have and whether you can use that knowledge to solve problems.

The Standard Grade Physics exam

The Standard Grade Physics exam is offered at two levels: General Level and Credit Level.

The General Level exam lasts for 1 hour 30 minutes and is the easier of the two papers. It has different types of question so you may have to:

- select the correct response from five options (multiple choice)
- write a short statement
- write a few sentences by way of a description or an explanation
- do a calculation.

About 40 marks are allocated to questions that test your knowledge of Physics and about the same number of marks are allocated to questions that test your ability to solve problems.

The Credit Level exam lasts for 1 hour 45 minutes. In this paper about 50 marks are allocated to questions testing your knowledge of Physics and about 50 marks are allocated to questions testing your ability to solve problems. The type of questions will be similar to General Level, but there aren't any multiple choice questions.

Whichever paper you have, you must answer all the questions in it – you don't have to choose from a selection. Both papers will cover the seven units of the Standard Grade Physics course. So expect questions on telecommunications, electricity, health physics, electronics, transport, energy and space physics.

In both papers, you are awarded a grade for knowledge and understanding and a grade for problem solving. In order to get the top grade in a paper, you will need to score 70% of the marks that are allocated.

In the exam

Using this book will help you to prepare for your exam. It is also worthwhile thinking ahead about your tactics in the exam. You can earn marks during the exam just by acting sensibly. Here are ten tips on examination technique.

Tip 1
Remember that every question in the paper is linked to something you have done already in the course. The examiner will not set a trick question. Questions will be asked in a way that makes it clear what you have to do. Everyone, including the examiner, wants you to do well.

Tip 2
If you cannot understand a question when you have read it, don't panic. Read it again more slowly and look carefully at any diagrams given in the question. Diagrams are there to help you.

Tip 3
When you come to the end of a numerical answer, make sure you include the correct unit.

Tip 4
Watch your timing as you work through the paper. At Credit Level, you earn 100 marks in 105 minutes. At General Level, you earn 80 marks in 90 minutes. This works out at about 1 mark a minute. So spend about 10 minutes on a 10-mark question.

Tip 5
When you come to the end of the exam paper, make sure you have answered all parts of the question and included the units for all your answers.

Tip 6
Where the wording of the question includes a phrase such as 'justify your answer', remember that a reason for your answer is needed. An answer on its own is not enough. You will not be awarded marks for guesswork.

Tip 7
Read the questions carefully. Ask yourself: 'What does the examiner expect of me?'; 'What is the question asking me to do?'; 'Do I have to calculate?'; 'Do I have to describe?'; 'Do I have to explain?' Answer what is asked. This seems like common sense, but in the heat of the moment this basic advice is often forgotten. Make your descriptions and explanations as clear as you can. Think about the important points you want to draw to the examiner's attention.

Tip 8
Remember that marks are awarded for partial success. You might be able to see part of the way to solve the problem. Show this to the examiner. However, don't hedge your bets by writing down everything related to the question asked. Examiners aren't fooled by this and you will only be wasting precious time. You need to show that you can identify and use what is relevant to solve a problem.

Tip 9
Look at the marks allocated to each question. Use them as a guide to how many statements you are expected to make. For example, if two marks are available, it is unlikely that you will gain full marks by making only one statement.

Tip 10
When you enter the examination room, try to stay calm. If you are well-prepared, there is no reason to panic. If you have worked hard on your revision, you will have given yourself every chance to perform to the best of your ability in the exam.

Good luck!

Forces and motion

This section is about:

- using correctly the key words (in bold) below
- describing motion using the terms 'average speed', 'speed' and 'acceleration'
- using speed-time graphs to describe motion
- explaining motion in terms of forces and using Newton's Laws
- using the terms 'weight' and 'gravitational field strength'
- explaining projectile motion and the motion of orbiting satellites.

To describe clearly how things move, you need to choose your words carefully. The words 'fast', 'slow', 'speeding up quickly' and 'slowing down sharply' give us an idea of what is happening. However, 'fast' to one person can mean 'slow' to another. The words 'quickly' and 'sharply' are vague. You need to use terms such as **speed** and **acceleration**, which have clear definitions, so that the words mean the same to everyone.

What kind of motion do you experience if you dive from a springboard? What type of motion is involved in jumping and throwing? These activities are examples of movement in a **gravitational field** where the **acceleration due to gravity** may be involved.

How is movement in space possible when there is nothing to push against? **Newton's Laws of Motion** enable us to answer questions like these.

Newton's Laws are about the cause of motion. What causes an apple to fall to the ground? You can explain falling in terms of the **weight** of an object or the **gravitational pull** acting on it. Weight is a **force**. When the force acting on an object is **unbalanced**, the object accelerates. Acceleration is the rate at which the speed of an object is changing.

When a car comes to a stop at traffic lights, its speed changes. It is slowing down or **decelerating**. This is the clue that an unbalanced force is at work – in this case, the **force of friction**.

Force, **mass** and acceleration are related to each other. We use a mathematical equation to describe the relationship. The equation is **force = mass × acceleration** or $F = ma$. This is **Newton's Second Law of Motion**.

FactZONE

Definition of speed and acceleration

$$\text{speed} = \frac{\text{distance travelled}}{\text{time taken}}$$

$$v = \frac{s}{t}$$

unit of speed is metres per second (m/s)

$$\text{acceleration} = \frac{\text{change in speed}}{\text{time taken for change}}$$

$$a = \frac{\Delta v}{t}$$

unit of acceleration is metres per second per second (m/s^2)

Speed-time graphs

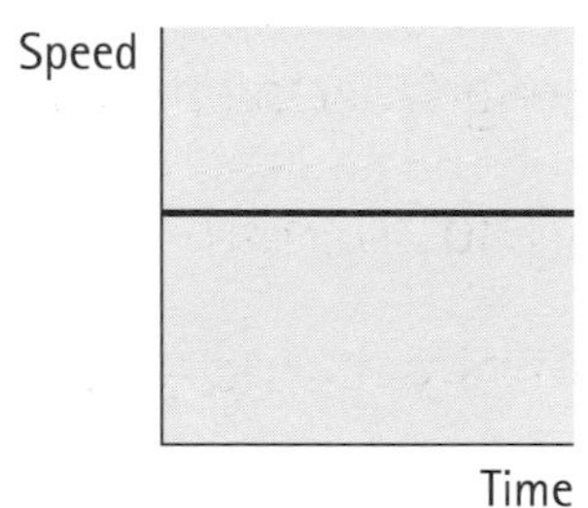

Constant speed

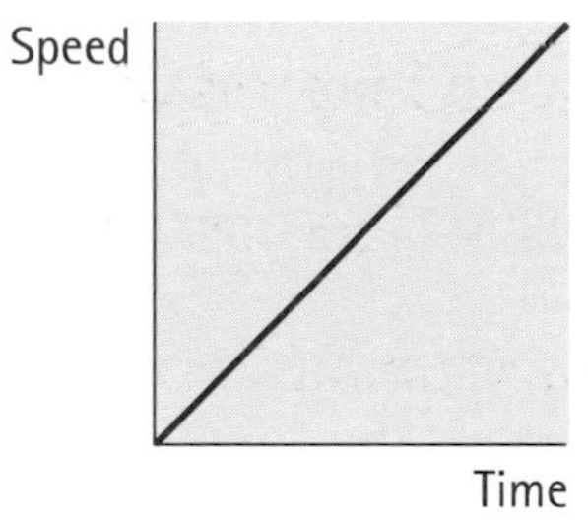

Uniform acceleration

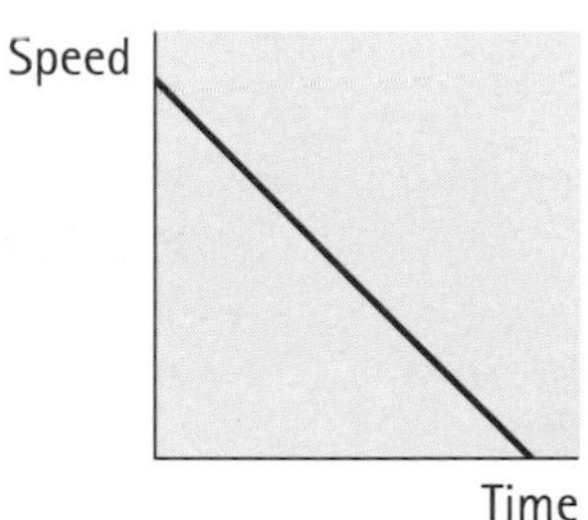

Constant deceleration

The area under a speed-time graph = distance travelled

Mass and weight

The mass of an object is the quantity of matter in the object. It is measured in kilograms (kg).

Weight is the gravitational pull on an object. It is a force and is measured in newtons (N).

weight = mass × gravitational field strength

$$W = mg$$

The force of friction

Friction is a force that acts between two surfaces in contact. It can oppose motion (e.g. brake pad rubs on the rim of a bicycle wheel to slow it down) or enable motion to take place (e.g. the friction force between the sole of your shoe and the pavement provides a force that enables you to move forwards).

Newton's First Law

Balanced forces acting on a body produce the same type of motion as no forces acting.

If the object is at rest and balanced forces are acting, the object remains at rest. If the object is in motion, it will move with a constant speed in a straight line.

Newton's Second Law

Unbalanced forces acting on an object produce an acceleration.

unbalanced force = mass × acceleration

$$F = ma$$

F is measured in newtons (N), m in kilograms (kg), a in metres per second per second (m/s^2)

Newton's Third Law

Forces never occur singly, always in pairs. If A pushes on B, then B pushes on A.

When you are walking, you use the force of friction to push backwards on the pavement. The pavement pushes forwards on you.

Rockets push out hot gases. The gases push back on the rocket enabling it to accelerate.

Speed and acceleration

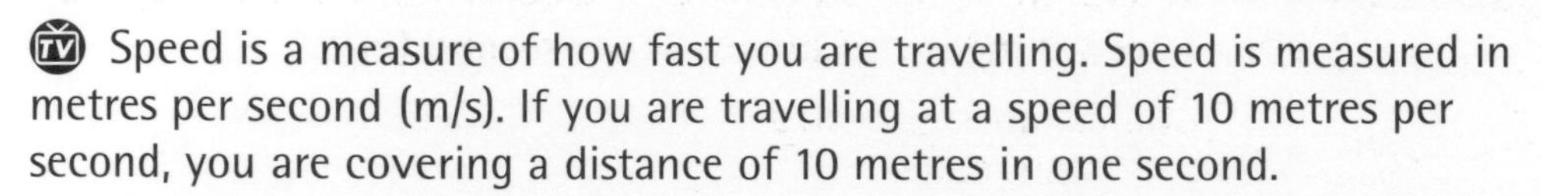

Speed is a measure of how fast you are travelling. Speed is measured in metres per second (m/s). If you are travelling at a speed of 10 metres per second, you are covering a distance of 10 metres in one second.

Acceleration is a measure of how quickly you can speed up (or slow down). It is measured in metres per second per second (m/s^2). An acceleration of 5 metres per second per second means that your speed is changing by 5 metres per second every second.

In a 100 m race, the sprinters accelerate from the starting blocks and reach their top speed in a short time. In sprinting, both speed and acceleration are important. The sprinter with the greatest top speed and the greatest acceleration will win the race.

In a Formula 1 race, where is a racing car's acceleration likely to be more important than its top speed?

Where in the race is the racing car's top speed likely to be more important than its acceleration?

Instantaneous speed

This is the speed of an object at a particular moment in time. Instantaneous speed is measured in metres per second (m/s).

The reading on the speedometer of a car at a particular time gives the instantaneous speed of the car. The driver uses the speedometer to check that the speed of the car is within the legal speed limit. Instantaneous speed is calculated by measuring the distance travelled in a short time interval and using the relationship:

$$\text{speed} = \frac{\text{distance travelled}}{\text{time taken}}$$

A toy car travels down a sloping runway. A card of length 5 cm attached to the roof of the car interrupts a light beam at the bottom of the runway. The time of the interruption is recorded as 0.1 s. Complete this calculation to find the instantaneous speed of the car in metres per second at the bottom of the runway.

$$\text{speed} = \frac{\text{distance travelled in metres}}{\text{time taken in seconds}}$$

$$= \frac{0.05}{\quad}$$

$$= \boxed{\quad}\ \text{m/s}$$

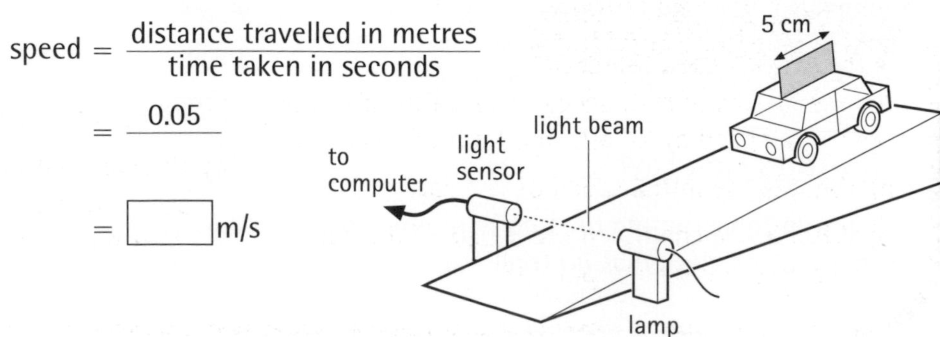

REMEMBER You are expected to use the terms 'constant speed' (or uniform speed), 'constant acceleration' (or uniform acceleration) and 'deceleration' in doing calculations on motion and when describing different types of motion.

Average speed

Some cars have a computer to record the time taken for a journey and the total distance travelled. It uses these measurements to calculate the average speed on a journey. Average speed is measured in metres per second (m/s).

$$\text{average speed} = \frac{\text{total distance travelled}}{\text{time taken}}$$

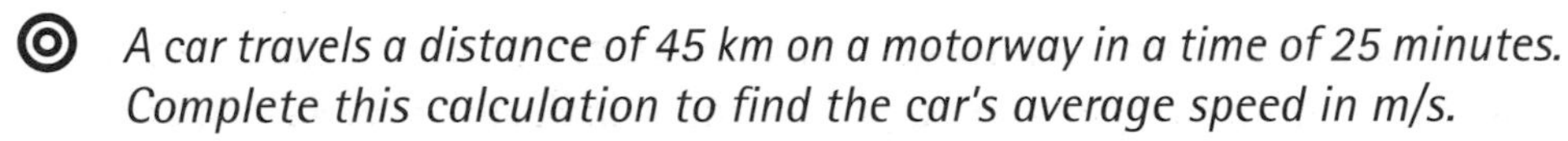

A car travels a distance of 45 km on a motorway in a time of 25 minutes. Complete this calculation to find the car's average speed in m/s.

$$\text{average speed} = \frac{\text{distance travelled in metres}}{\text{time taken in seconds}}$$

$$= \frac{45 \times \quad}{25 \times \quad}$$

$$= \square \text{ m/s}$$

If a speed of 1 m/s corresponds to 2.2 miles per hour (mi/h), what is the car's average speed on the journey in mi/h? Is it possible to say if the car exceeded the speed limit of 70 mi/h on the journey? Explain your answer.

Acceleration

Acceleration is a measure of how quickly something is speeding up or slowing down. Acceleration is the rate at which the speed is changing.

$$\text{acceleration} = \frac{\text{change in speed}}{\text{time taken}}$$

$$= \frac{(\text{final speed} - \text{initial speed})}{\text{time taken}}$$

Acceleration is measured in metres per second per second (m/s^2).

A runner starts from rest (0 m/s) and reaches a speed of 9 m/s in a time of 4 s. Complete this calculation to find the acceleration of the runner.

$$\text{acceleration} = \frac{\text{change in speed}}{\text{time taken}}$$

$$= \frac{(9 - 0)}{\quad}$$

$$= \square \text{ m/s}^2$$

REMEMBER We sometimes say that a moving object has a 'constant acceleration' or a 'uniform acceleration'. This means the speed of the object changes by the same amount every second. When an object's speed is decreasing with time (i.e. slowing down), its speed is changing and so, by definition, the object is accelerating. However, we often refer to this type of motion as a 'deceleration'.

Practice question

A car starts from rest, accelerates uniformly and reaches a speed of 26 m/s in a time of 8 seconds.

a) What is the initial speed of the car?

b) What is the change in the speed of the car during the 8 second period?

c) What is the acceleration of the car?

d) The car continues with the same acceleration. What is the speed of the car after another 2 seconds?

Speed–time graphs

The terms 'speed', 'constant speed', 'average speed' and 'uniform acceleration' describe different types of motion. Graphs can also describe motion. These graphs are called speed-time graphs.

In Standard Grade Physics you have to know how to draw and interpret speed-time graphs that represent situations involving constant speed and uniform acceleration.

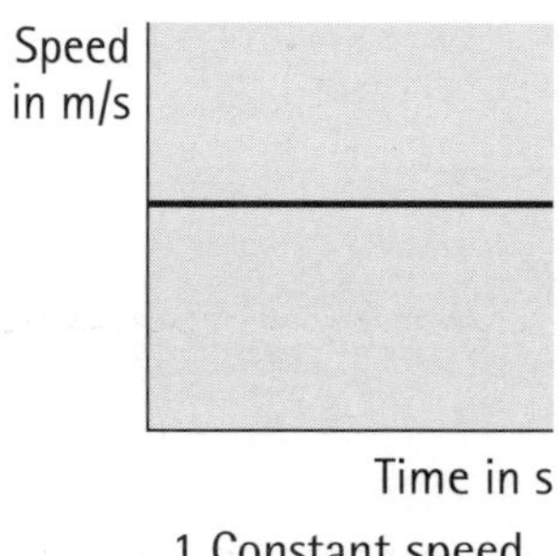

1 Constant speed

Speed
in m/s

Time in s

2 Uniform acceleration

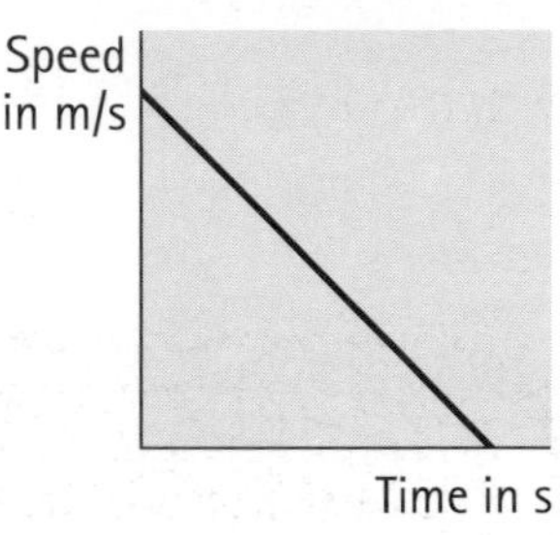

3 Uniform deceleration

REMEMBER A uniform speed is one that does not change over time. A uniform acceleration is one where the speed changes by the same amount each second.

1) A horizontal line on a speed-time graph represents a constant speed.

2) A sloping line on a speed-time graph represents an acceleration. This sloping line shows that the speed of the object is changing. Here it is speeding up. The steeper the slope of the line, the bigger the acceleration.

3) If the line slopes downwards from left to right on the graph, like this, the object is slowing down. This motion is sometimes called a 'deceleration'.

This is a speed-time graph of a bus.

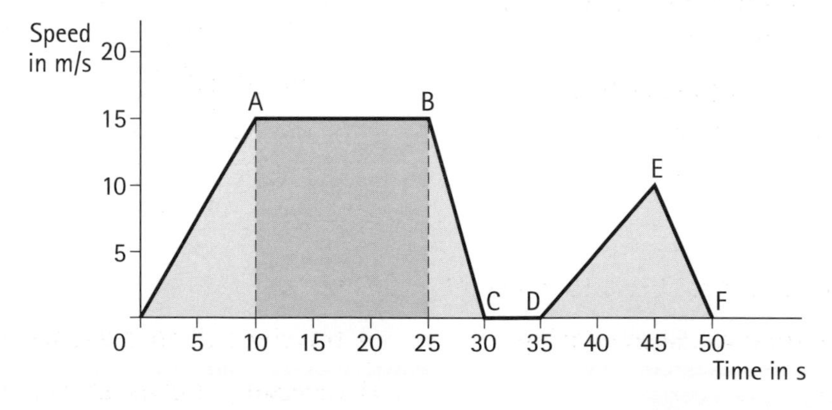

During parts OA, BC, DE and EF the bus has a 'uniform acceleration'. Parts BC and EF could also be described as 'uniform deceleration'. The bus has a 'uniform speed' during part BC and is 'at rest' during part CD.

$$\text{The acceleration of the bus in part OA} = \frac{\text{change in speed}}{\text{time taken}}$$

$$= \frac{15}{10}$$

$$= 1.5\ \text{m/s}^2$$

The deceleration of the bus during part BC $= \dfrac{\text{change in speed}}{\text{time taken}}$

$= \dfrac{15}{5}$

$= 3\ \text{m/s}^2$

◎ *Show that the acceleration of the bus during part DE is 1 m/s² and that the deceleration during part EF is 2 m/s².*

The distance travelled by the bus is the area under the speed-time graph. You can calculate this area by dividing the graph up into rectangular and triangular areas.

distance travelled $= (\tfrac{1}{2} \times 10 \times 15) + (15 \times 15) + (\tfrac{1}{2} \times 5 \times 15) + (\tfrac{1}{2} \times 15 \times 10)$

$= 75 + 225 + 37.5 + 75 = 412.5\ \text{m}$

Practice questions

1) Passengers in a car record the speed of the car at different times during a short journey. The speeds recorded at different times are shown in the table.

Use the information in the table to complete the speed-time graph.

Time in s	0	5	10	15	20	25	30	35
Speed in m/s	0	7.5	15	22.5	22.5	22.5	11.25	0

Now look at the graph that you have drawn.

a) What is the car's top speed?

b) Calculate the car's acceleration and deceleration.

Ⓒ c) How far does the car travel during the journey?

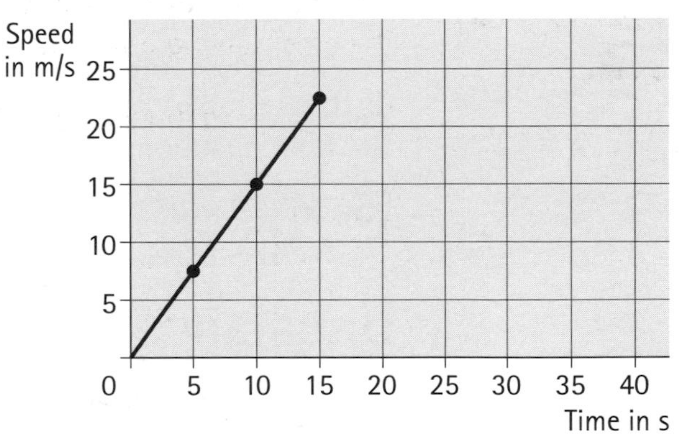

2) Which graph is most likely to represent a car that is:

a) accelerating whilst overtaking

b) braking

c) moving off from the traffic lights

d) travelling at constant speed on the motorway

e) reversing at constant speed?

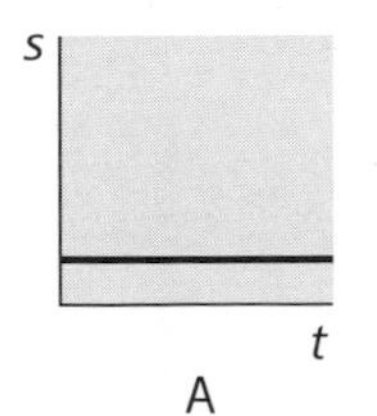

A

s
t

B

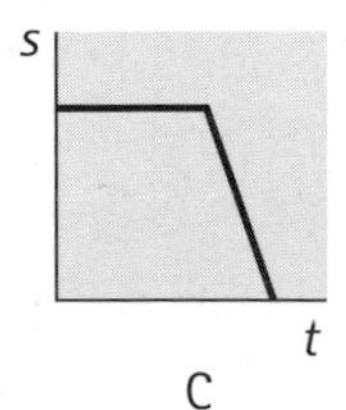

C

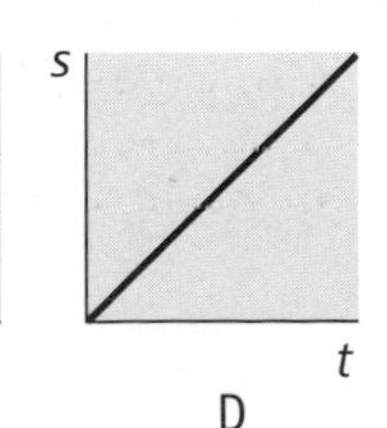

D

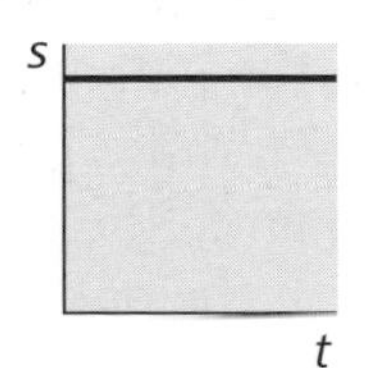

E

s = speed
t = time

Forces and Newton's Laws

REMEMBER The forces acting on an object cause the motion of the object to be either changing or unchanging.

Changing and unchanging motion

The motion of an object is either 'changing' or 'unchanging'.

Unchanging motion is when the object is at rest or is moving with a steady speed in a straight line.

Changing motion includes movement where the object is speeding up or slowing down. It also includes motion where the direction in which the object moves is changing. An object that is moving in a curved path is an example of changing motion.

Balanced forces

Balanced forces are responsible for unchanging motion.

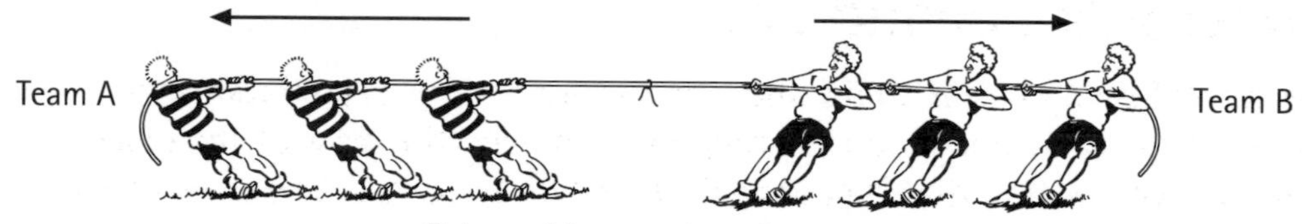

Balanced forces act on the rope

This tug of war, where each team is pulling equally on the rope, is an example of unchanging motion. Both teams are pulling on the rope, but the force due to one team is cancelled or balanced by the force due to the other. Balanced forces are forces where the effect of one force is cancelled out by another. The rope is stationary under the action of these balanced forces.

Which of the forces shown in the diagram is produced by team A?

Unbalanced forces

Unbalanced forces cause 'changing motion'.

The lift-off of the space shuttle is an example of changing motion.

The space shuttle accelerates upwards from its launch pad. The rocket is acted on by two forces. One is a downward pulling force, the other is an upward pushing force. The pulling force is caused by the pull of gravity, i.e. the weight of the rocket. The upward pushing force comes from the rocket engines. The push from the engines is greater than the pull of gravity, so the forces are unbalanced – they do not cancel each other out. These unbalanced forces cause acceleration.

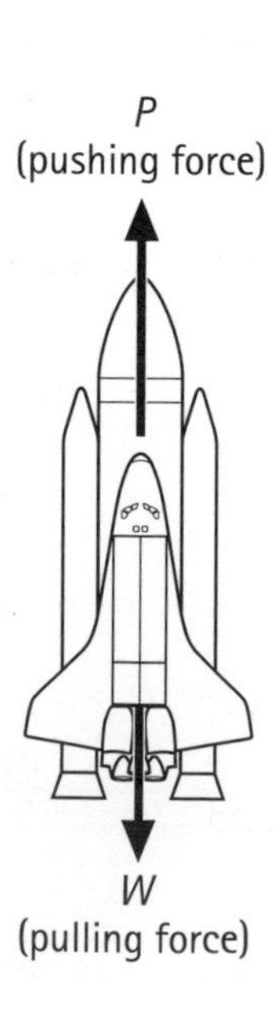

There are two points to note about the acceleration of an object when an unbalanced force acts on it.

- The bigger the unbalanced force acting on the object, the bigger the acceleration of the object.
- The more mass the object has, the more inclined it is to resist any change to its motion. For example, if you apply the same unbalanced force to a mass of 1000 kg and a mass of 1 kg, the acceleration (change in motion) of the 1000 kg mass will be much less than that of the 1 kg mass.

Newton's First Law

Newton's First Law is about balanced forces. It states that if a body is at rest and the forces acting on it are balanced, the body will remain at rest. However, if the body is moving and the forces acting are balanced, the body will keep moving at constant speed in a straight line.

REMEMBER Balanced forces give no motion or motion in a straight line with constant speed.

Newton's Second Law

Newton's Second Law is about unbalanced forces. It gives a relationship between the unbalanced force acting on the object, the mass of the object and the acceleration that is produced. The relationship can be expressed by the following straightforward mathematical equation.

REMEMBER Unbalanced forces produce an acceleration.

unbalanced force = mass × acceleration

$$F = ma$$

The unbalanced force F is measured in newtons (N).

The mass m is measured in kilograms (kg).

The acceleration a is measured in metres per second per second (m/s^2).

As part of your Standard Grade Physics course, you are expected to be able to do calculations involving Newton's Second Law. At General Level, you need to know how to find the value for the quantities F or m or a when you are given information about the other two of the quantities.

Write the above equation in its three forms.

F = ... × ....

a = ... / ....

m = ... / ....

At Credit Level, as well as doing calculations like these, you might first have to work out what the unbalanced force is before you use Newton's Second Law.

Newton's Third Law

When a rocket accelerates upwards from its launch pad, the upward push on the rocket is a consequence of Newton's Third Law of Motion.

A simple way of stating this Law is to say:

'If A exerts a force on B, then B exerts a force on A.'

At Credit Level, you have to be more precise in your statement of the Law and say:

'If A exerts a force on B, then B will exert an equal and opposite force on A.'

REMEMBER Weight is another name for the pull or the force of gravity acting on an object. Weight is measured in newtons (N).

Newton's Third Law states that when you push on something, that something – whatever it is – pushes back on you.

You can explain rocket motion in terms of Newton's Third Law.

At lift-off, hot burning gas is pushed downwards by the rocket motors. The hot gas pushes back on the rocket in an upward direction. When this upward push, or thrust, exceeds the weight of the rocket, the forces acting on the rocket are unbalanced and the rocket accelerates upwards.

You can also use Newton's Third Law to explain how astronauts are able to 'walk' in space.

Astronauts in space can manoeuvre using a special back-pack. The back-pack has small jets which eject nitrogen gas. When gas is being ejected from the jets, the gas pushes back on the astronaut allowing him or her to move in different directions in space.

Here is another example of Newton's Third Law at work.

This balloon is taped to a straw. The straw is free to move along the length of tight, smooth thread. The balloon is held at the neck so that air does not escape from it.

The balloon is released and it accelerates along the thread. The cause of the movement of the balloon along the thread is similar to the cause of the motion of a rocket at lift-off.

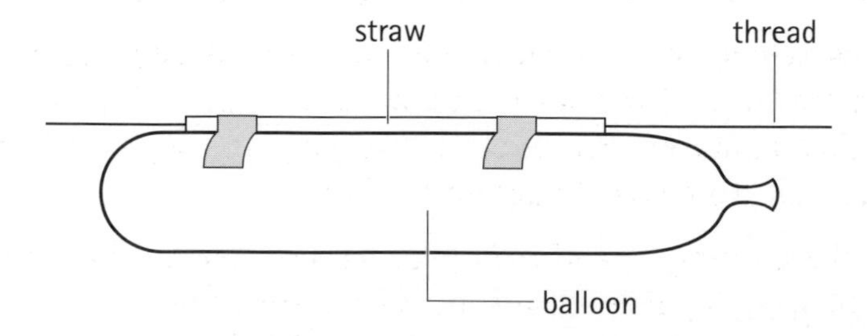

What is the force exerted by 'A' and what is the force exerted by 'B'?

On the diagram, show the direction of the force that acts on the balloon.

Practice questions

1) A car has a mass of 1200 kg. An unbalanced force of 2400 N acts on the car. What is the acceleration of the car?

2) Three horizontal forces are applied to a trolley of mass 4 kg as shown in the diagram.

C a) What is the unbalanced force acting on the trolley?

C b) What is the acceleration of the trolley?

3) A spaceship of mass 15 000 kg is at rest in deep space. The spaceship has four small identical thruster rockets, P, Q, R and S, which allow the spaceship to be manoeuvred. The spaceship is moved to the right by firing two of the thruster rockets.

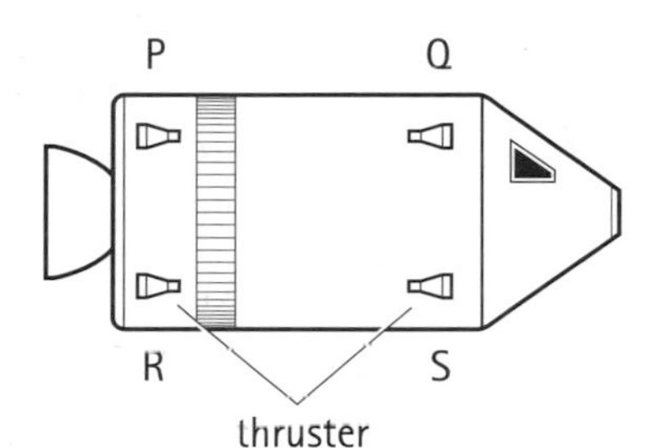

After 10 seconds, the speed of the spaceship is 2 m/s. The thruster rockets are then switched off.

a) Which two of the thruster rockets were fired?

b) What is the acceleration of the spaceship when the thruster rockets are being fired?

c) What is the thrust provided by one thruster rocket?

d) What is the speed of the spaceship after the thruster rockets are switched off?

e) Which thruster rockets should be fired to bring the spaceship to rest?

4) Newton's First and Second Laws can be used to explain why it is important to wear seatbelts. Complete the paragraphs below using the following words.

speed *force* *balanced* *deceleration* *unbalanced*

By Newton's First Law, when a car is travelling at constant speed in a straight line, the forces acting on the car and the passengers are .... If the car has to brake sharply to avoid a collision, its motion is changing, and so, by Newton's Second Law, an .... force has to act on the car.

It is important that the passengers and the driver reduce their speed at the same rate as the car, i.e. the passengers and car should have the same .... If this does not happen, the passengers will move forwards with a greater .... than the car and perhaps collide with the windscreen or with each other. The wearing of seatbelts enables a .... to be applied to the passengers as the car is braking. The force from the seatbelts ensures that the passengers reduce their speed at the same rate as the car and the risk of injury is reduced.

Weight and gravity

The force of gravity

The Earth has a gravitational field. Objects in the Earth's gravitational field experience a gravitational pull. The gravitational field at the surface of the Earth produces a force of approximately 10 N on every mass of 1 kg. We say that the Earth has a gravitational field strength of 10 N/kg.

The gravitational field strength of the Earth is shown by the symbol g.

$g = 10$ N/kg

The gravitational field strength gets weaker as you move further away from the surface of the Earth.

The pull that the Earth's gravitational field has on an object is called the 'force of gravity'.

The force of gravity acting on an object is usually called the 'weight' of the object.

You can calculate the weight of an object by using the relationship:

$W = mg$

W is the weight of the object in newtons (N).
m is the mass of the object in kilograms (kg).
g is the gravitational field strength in newtons/kilogram (N/kg).

The strength of the gravitational field on the Moon is much less than the strength of the gravitational field of the Earth. On the Moon, objects weigh about $\frac{1}{6}$ of what they weigh on Earth.

What is the gravitational field strength of the Moon?

Acceleration due to gravity

REMEMBER Mass and weight are different quantities. Mass is a measure of the quantity of matter in an object and it is measured in kilograms (kg). Weight is a force. It is the gravitational pull on an object and is measured in newtons (N).

If the only force acting on an object is its weight, there is an unbalanced force acting on the object. The object will accelerate and fall freely. The acceleration of an object which is falling freely is called the 'acceleration due to gravity'.

You can work out the value of the acceleration due to gravity by applying Newton's Second Law to the motion of a freely falling diver.

John has a mass of 50 kg and is standing on the edge of a high diving board. He steps off the board and falls freely to the pool below.

◎ *Show the forces acting on John when he is standing on the board.*

The unbalanced force acting on John when he steps off the board is W.

W = mass × gravitational field strength

= 50 × 10

= 500 N

You can now apply Newton's Second Law to the motion.

unbalanced force = mass × acceleration

500 = 50 × acceleration

$$\text{acceleration} = \frac{500}{50}$$

$$= 10 \text{ m/s}^2$$

So the acceleration of John falling freely = 10 m/s^2

REMEMBER N stands for newtons and m/s^2 stands for metres per second per second.

The value of 10 m/s^2 that you calculated for John is called the 'acceleration due to gravity'. The acceleration due to gravity on the Earth is the same for all objects. This may seem strange because everyone knows that if you drop a feather and a coin at the same time, the coin will reach the floor first.

? *Explain the difference in the motion of the feather and the coin in terms of air resistance.*

If you were to drop objects where there was no atmosphere and, therefore, no air resistance, you would get a quite different result. For example, an astronaut on the Moon dropped a hammer and a feather. They fell with the same acceleration.

Practice questions

1) A car of mass 750 kg rests on a hydraulic ramp in a garage. The ramp applies a force P to the car and raises it at a steady speed.

   a) What is the weight of the car?

   b) What is the size of force P?

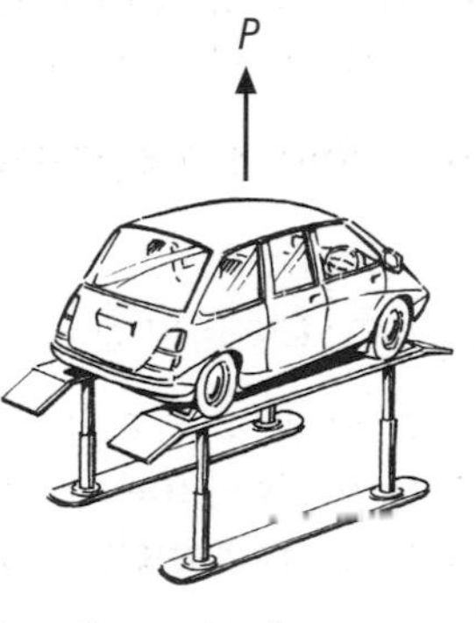

2) Two forces act on a parachutist during a drop. One force is the weight W of the parachutist. The other force is the upward pull P from the parachute harness on the parachutist. The parachutist has a mass of 60 kg.

   a) What is the weight of the parachutist?

   b) What is the size and direction of the unbalanced force acting on the parachutist?

   c) What is the acceleration of the parachutist?

Projectiles, satellites, weightlessness

Projectiles

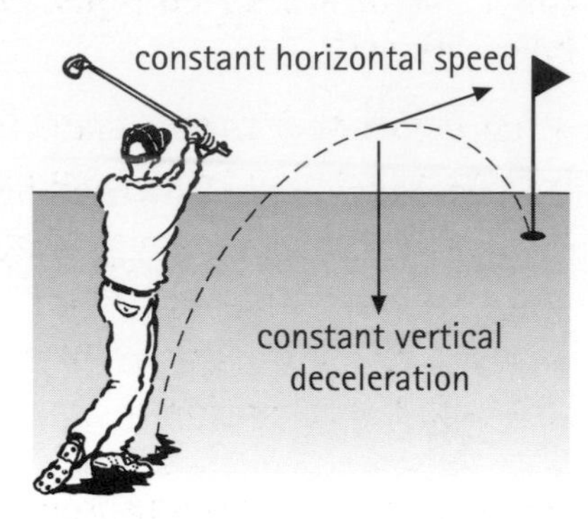

When a discus is launched, a javelin is thrown, a golf ball is hit into the air or a tennis ball is served, the movement that takes place is an example of projectile motion. High jumping, long jumping and hurdling are also examples of projectile motion. If you ignore air resistance, the only force that acts on a projectile during its motion is its weight.

Projectile motion is a combination of two independent motions:

- constant horizontal speed
- constant vertical acceleration.

REMEMBER At Credit Level, you need to know how to calculate the motion of projectiles.

As the projectile moves outwards, it is being pulled downwards by the force of gravity. The combination of these two motions gives rise to a curved flight path.

You can explain projectile motion in terms of Newton's First and Second Laws.

A ball is projected horizontally from a table top. If you ignore air resistance, there are no forces acting on the ball in a horizontal direction. So its motion in this direction follows Newton's First Law. The horizontal speed of the ball does not change.

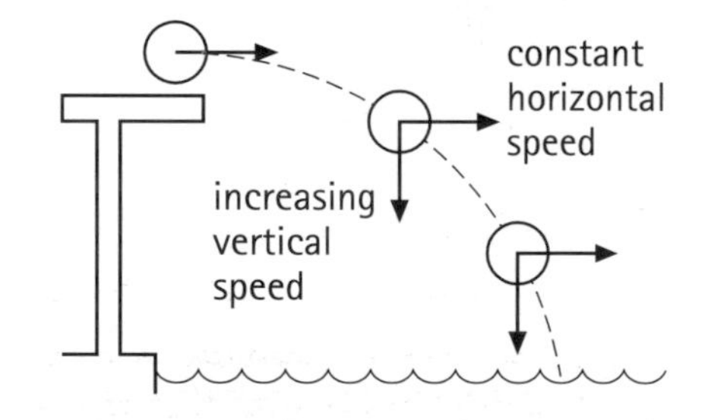

The horizontal motion can be shown in a speed-time graph.

The area under this graph gives the distance that the ball has travelled horizontally.

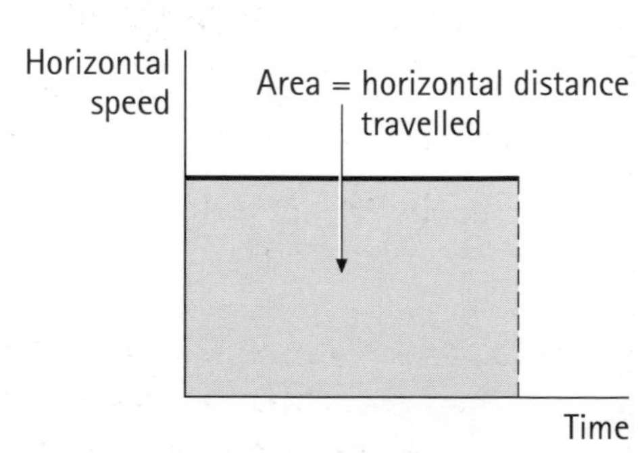

In the vertical direction, there is a force acting. This is the weight of the ball. So the vertical motion follows Newton's Second Law. The ball falls downwards with a constant acceleration – the acceleration due to gravity. The vertical motion can also be shown in a speed-time graph.

The initial downward speed of the projectile is zero. The area under this graph gives the vertical distance dropped by the projectile.

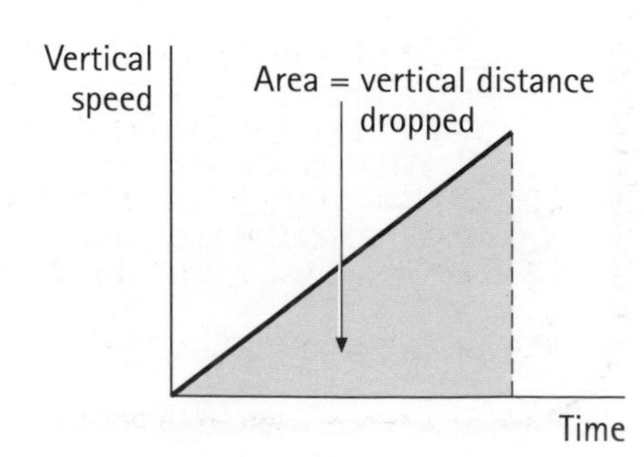

Satellites and weightlessness

Satellite motion is an example of projectile motion. A satellite or a spacecraft in orbit round the Earth is actually falling freely towards the centre of the Earth. The orbital speed is just enough to make it keep missing the Earth as it falls towards it.

A spacecraft and its occupants in orbit around the Earth are in free fall. The spacecraft and its occupants all have the same acceleration.

Objects in the spacecraft look like they are floating as though they are weightless. The objects are not really weightless. The appearance of weightlessness is because all objects in the spacecraft have the same downward acceleration.

REMEMBER When you are in free fall, your acceleration is the acceleration due to gravity and the only force acting on you is your weight.

Imagine you are stepping from a diving board whilst holding a ball.

If you let go of the ball on the way down, how does the ball move?

The ball does not fall downwards relative to you. It stays alongside you during your descent. This is because both you and the ball have the same downward acceleration. The ball appears to float relative to you – it looks like it is weightless.

If you gave the ball a sideways push during your descent, how would the ball move relative to you?

Practice questions

1) a) In the diagram above, sketch the path followed by the ball (after it is pushed) as seen by you when you are falling freely from the diving board.

   b) Sketch the path followed by the ball (after it is pushed) as seen by the observer at the poolside.

2) A stone of mass 0.25 kg is thrown horizontally with a speed of 7 m/s from the top of a vertical cliff. The stone is in the air for 2 seconds before it hits the beach below. Ignore the effect of air resistance on the motion of the stone.

   a) What is weight of the stone?

   b) What is the acceleration of the stone?

   c) How far does the stone land from the base of the cliff?

   d) Sketch the speed-time graph for the stone's vertical motion.

   e) How high is the cliff?

Examination questions

Try these two questions from past exam papers. Question 1 is from a **General Level** paper and question 2 is from a **Credit Level** paper. Spend about 7 minutes on question 1 and no more than 12 minutes on question 2. When you have finished, turn to page 94 for the answers and the marking guide.

1) The table below gives some information about the performance of three cars A, B and C.

Car	**A**	**B**	**C**
Top speed in miles per hour	116	116	122
Time in seconds to accelerate from 0 to 60 miles per hour	12	10	10
Acceleration (from 0 to 60 miles per hour) in miles per hour per second	-----	6	6

a) The value for the acceleration of car A as it travels from 0 to 60 miles per hour has not been entered into the table. Calculate the missing acceleration for car A. **(2)**

b) One of the cars takes part in a trial run on a race track. The speed-time graph for the motion is shown below.

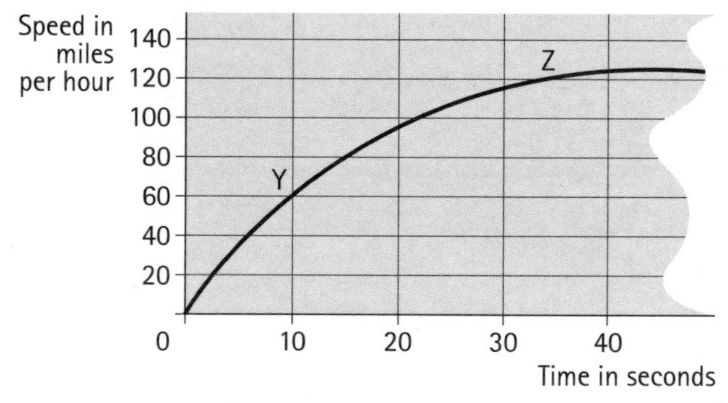

i) Identify the car whose motion is represented by the graph. Explain your choice. **(2)**

ii) Describe the motion of the car between Y and Z on the graph. **(1)**

c) The values given in the car performance table are for the cars carrying no passengers. What effect would carrying passengers have on the time taken to accelerate from 0 to 60 miles per hour? **(1)**

C 2) Competitors are taking part in a bobsleigh competition.

a) Figure 1 shows the bobsleigh at point X near the end of the run.

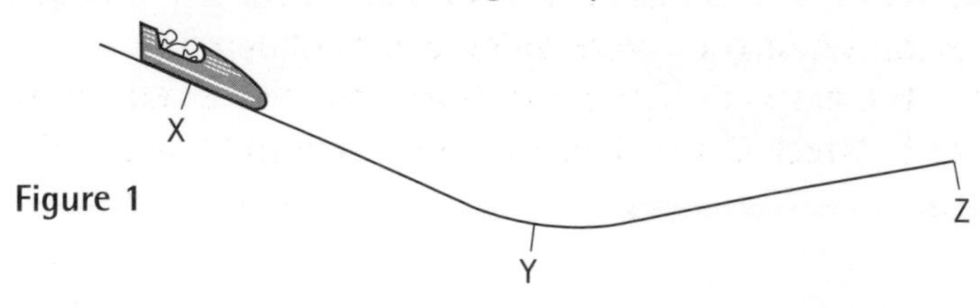

Figure 1

When the bobsleigh reaches point Y, the brakes are applied until it comes to rest at point Z. The speed-time graph of the motion of the bobsleigh from point X to point Z is shown in figure 2.

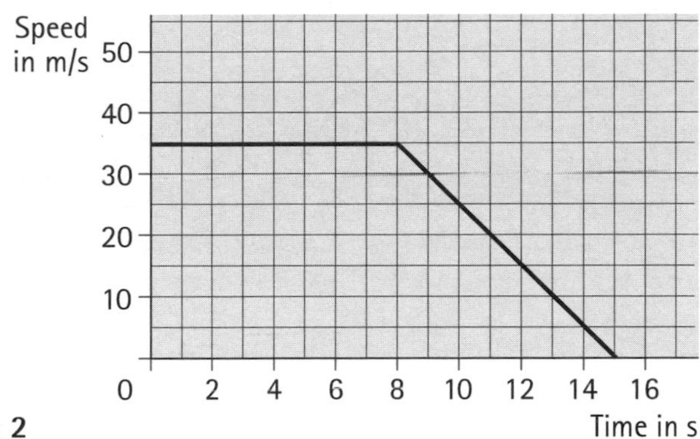

Figure 2

i) What time did the bobsleigh take to travel from X to Y? (1)
ii) What was the distance travelled by the bobsleigh from point X until it came to rest? (3)
iii) Calculate the deceleration of the bobsleigh between Y and Z. (2)
iv) The competitors and the bobsleigh have a total mass of 380 kg. Calculate the force causing the deceleration of the competitors and bobsleigh. (2)

b) At the start of the run, at point A in figure 3, both competitors push the empty bobsleigh. At point B, one of the competitors jumps in while the other keeps pushing. At point C, the second competitor jumps in.

Figure 3

The speed-time graph of the motion of the bobsleigh from A to C is shown in figure 4.

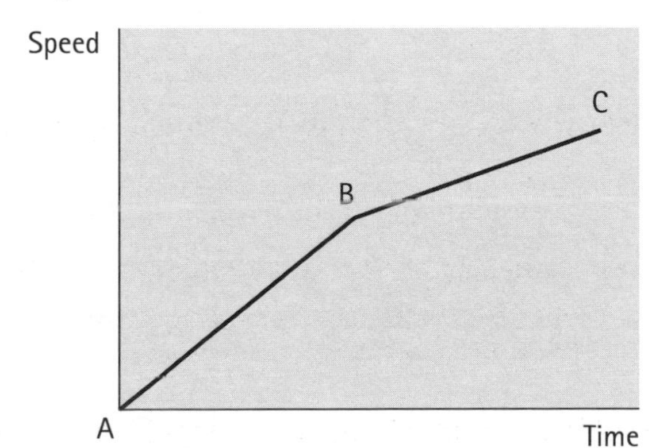

Figure 4

i) Give **two** reasons, in terms of Newton's Laws, for the change in the acceleration of the bobsleigh between AB and BC. (2)
ii) Complete the graph in figure 4 to show how the speed varies with time between C and D when both competitors are in the bobsleigh. (1)

Energy

This section is about:

- using correctly the key words (in bold) below
- describing motion in terms of kinetic energy and gravitational potential energy
- solving problems involving work, power, energy transfer and the law of conservation of energy
- describing methods of transferring heat and carrying out calculations involving specific heat capacity and latent heat.

Energy is needed to make things work. Energy in the form of fuel is needed to make a car engine work. A car engine transfers the **chemical energy** in the fuel to energy associated with movement – **kinetic energy**. Your body is a kind of engine. Energy in the form of food is needed to make your body work. If you attach this engine to a bicycle, you can transfer chemical energy from your body to kinetic energy.

When you work out at the gym, you transfer energy. When energy is being transferred from one form to another, **work** is being done. When you apply a **force** and move that force over a **distance**, you have done work.

The work you do in the gym can be calculated automatically for you by electronic circuits which are built into the exercise machines. The circuits can also tell you the rate at which you have been working. This is known as your **power**.

An important law in Physics is the **Law of Conservation of Energy**. This law says that you cannot use up energy. All you can do is transfer it from one form to another. For example, you can apply this law to the motion of a person on a swing. As the person swings to and fro, energy is transferred from **gravitational potential energy** to kinetic energy and vice versa. In theory, the person will keep swinging to and fro forever. But, in practice, this does not happen. The person comes to rest after a number of swings have taken place. This is because during each swing some energy is transferred as **heat energy**.

Heat energy is used in power stations to produce the steam that drives **turbines** which, in turn, drive the generators that provide the **electricity** for industry and our homes. Heat energy raises the **temperature** of water in our houses.

Heat energy also raises the temperature of iron ore to produce molten metal from which steel is made. Here the heat energy is producing a **change of state** – a solid substance is changed to a liquid. The heat energy involved in a change of state is called **latent heat**.

FactZONE

Kinetic energy and gravitational potential energy

Kinetic energy (E_k) is the energy associated with the movement of an object. Gravitational potential energy (E_p) is the energy an object has because of its height (h) in a gravitational field.

G $E_k = \frac{1}{2} mv^2$

$E_p = mgh$ where g is the gravitational field strength measured in newtons per kilogram (N/kg)

Work and power

work done = force × distance = energy transferred

Work done is measured in joules (J).

Power is the rate of doing work.

$$\text{power} = \frac{\text{energy transferred}}{\text{time taken}}$$

G Law of Conservation of Energy

If you ignore the effect of frictional forces, you can describe, explain and predict motion in a gravitational field using two conservation of energy equations:

loss of kinetic energy = gain in gravitational potential energy

gain in kinetic energy = loss of gravitational potential energy

In practice, there is always some work done against friction and some energy is transferred as heat (E_h). So a more exact conservation of energy equation is:

original E_p + original E_k = final E_p + final E_k + E_h

Heat and specific heat capacity

The heat energy (E_h) absorbed by a body of mass (m) is calculated using the equation:

$E_h = cm\,\Delta T$

c is the specific heat capacity of the body measured in J/kg °C

ΔT is the change in temperature measured in °C

Change of state and latent heat

Supplying heat energy to a substance can change its state. But this does not involve a change in temperature.

G $E_h = ml$ applies to the changing of a mass m of substance in its solid state at its melting point to a liquid at the same temperature. Here the symbol l stands for the specific latent heat of fusion.

The heat energy (E_h) that is needed to change a mass (m) of liquid at its boiling point into vapour at the same temperature is calculated using $E_h = ml$

l is the specific latent heat of vaporisation measured in joules per kilogram (J/kg).

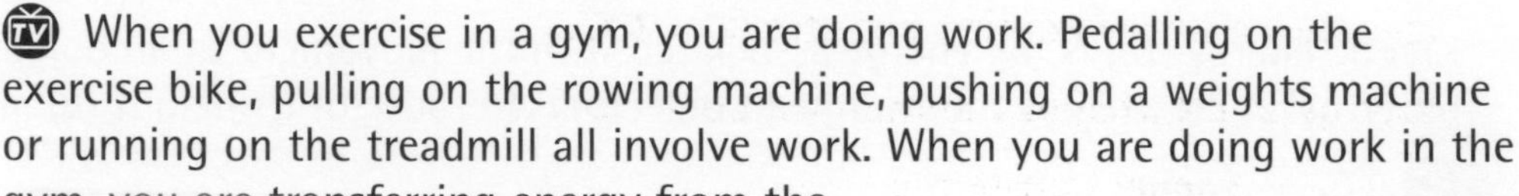

Work, energy and power

When you exercise in a gym, you are doing work. Pedalling on the exercise bike, pulling on the rowing machine, pushing on a weights machine or running on the treadmill all involve work. When you are doing work in the gym, you are transferring energy from the chemical energy which is stored in your body to other forms of energy. The work that you do may be transferred as kinetic energy as you pedal on the exercise bike. Or the work may involve transferring energy to gravitational potential energy. This is the energy that is given to a load when you lift it through a height.

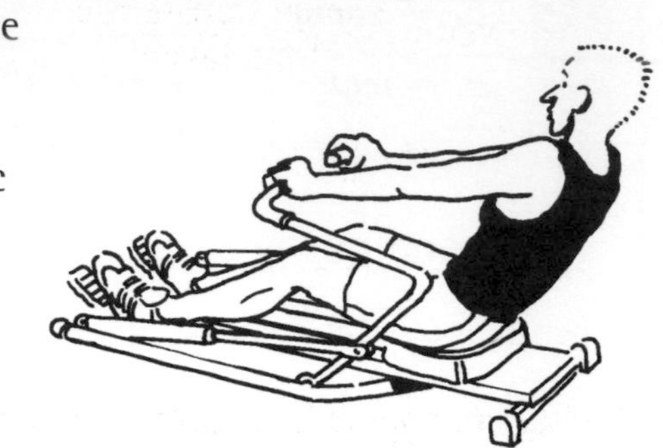

Work and energy transfer

REMEMBER N stands for newtons and J stands for joules.

When work is being done, energy is being transferred and a force is being applied over a distance. The amount of work that is done is measured by multiplying the force that is applied by the distance over which the force has been applied.

On the rowing machine, Helen pulls the oar with a force of 160 N through a distance of 1.25 m during one stroke. Complete this calculation to find out how much work is done during one stroke.

work done = force × distance

= .... × ....

= .... joules

The energy that Helen transferred during one stroke is 200 J and some body fat was burned off to provide this energy.

Power

REMEMBER When you are doing work, you are transferring energy. Your rate of doing work is called your 'power'.

Instead of how much work is done, we are often more interested in how quickly work can be done. Power is the rate of doing work or the rate at which energy is being transferred.

$$\text{power} = \frac{\text{work done}}{\text{time taken}}$$

$$= \frac{\text{energy transferred}}{\text{time taken}}$$

Power is measured in joules per second or watts (W).

When Helen is rowing, the energy transferred during one stroke is 200 J. Her stroke rate is 30 strokes per minute. Complete this calculation to show that Helen's power is 100 watts.

energy transferred during one stroke = 200 J

energy transferred in 1 minute = 200 × .... J

$$\text{power} = \frac{\text{energy transferred}}{\text{time taken}}$$

$$= \frac{\quad}{60}$$

= .... J/s

= 100 watts

REMEMBER Power is the rate at which energy is transferred. Power is measured in joules per second (J/s) or watts (W).

The human engine is capable of generating a few hundred watts of power. If you could harness this power, it could be used to generate electricity for a few 100 W lamps. Car engines are more powerful than the human engine. The engine of a small car can generate around 45 kW of power.

How many 100 W light bulbs deliver the same power as the car engine? How many 1-bar 1 kW electric fires deliver the same power as the car engine?

Practice questions

1) A load of mass 40 kg is raised 10 times through a distance of 0.8 m during a fitness exercise.

   a) What is the weight of the load?

   b) How much work is done during the whole exercise?

2) Liam rows a distance of 1500 m on a rowing machine in 7.5 minutes. The digital read-out on the rowing machine shows that the power developed during the row is 200 W.

   a) What is the average speed that is simulated by the row?

   b) How much energy is transferred by the machine during the row?

   c) Estimate the average force Liam exerts on the 'oars' during the exercise.

3) On a fitness bicycle, Gita pedals against a friction force of 50 N. The electronic system records that a distance 1.5 km is travelled in a time of 5 minutes.

   a) How much work is done by Gita in 5 minutes?

   b) How much energy is transferred by the bicycle during the 5 minutes?

   c) What is the power developed by Gita during the exercise?

Kinetic energy and gravitational potential energy

Kinetic energy

When you pedal your bicycle so that its speed increases, you are transferring chemical energy from your muscles to the kinetic energy of the bicycle. Kinetic energy is the energy an object has as a result of its movement. The amount of kinetic energy a moving object has depends on the mass of the object and its speed. The bigger the mass and the greater the speed of the object, the greater its kinetic energy.

Kinetic energy is given by the equation:

$$E_k = \frac{1}{2} mv^2$$

REMEMBER At Credit Level, you are expected to do calculations using the relationship: $E_k = \frac{1}{2} mv^2$

E_k is the kinetic energy of the object measured in joules (J).
m is mass of the moving object measured in kilograms (kg).
v is the speed in metres per second (m/s).

Complete the table comparing the kinetic energy of a golf ball on a putting green with a cyclist of mass 80 kg travelling at 10 m/s (22 miles per hour) and a car of mass 750 kg travelling at 30 m/s (66 miles per hour).

	Mass in kg	Speed in m/s	E_k in J
Golf ball	0.05	3	0.225
Cyclist			
Car			

REMEMBER Remember that gravitational field strength is the gravitational pull on a mass of 1 kg. It is measured in newtons/kilogram (N/kg).

Gravitational potential energy

When an object is raised above ground level, work is done on the object, and energy is transferred to the object in the form of gravitational potential energy.

Gravitational potential energy is the energy an object has because of its position above the surface of the Earth.

The gravitational potential energy (E_p) of an object of a given mass (m) at a height (h) is shown by the equation:

$$E_p = mgh$$

REMEMBER For positions near the surface of the Earth g = 10 newtons per kilogram (N/kg).

m is the mass of the object in kilograms (kg).
g is the gravitational field strength, i.e. the gravitational pull of Earth for a mass of 1 kg.
h is the height above the surface of the Earth.

The expression mgh for gravitational potential energy comes from:

work done = force × distance

Joshua raises a load of a given mass (m) through a height (h). The weight of the load is mg. In lifting the load, he has to pull upwards on it with a force of size mg. When this force is applied over the distance h then:

$$\text{work done} = \text{force } (mg) \times \text{distance } (h)$$
$$= mgh$$
$$= \text{gravitational potential energy of load}$$

Therefore $E_p = mgh$

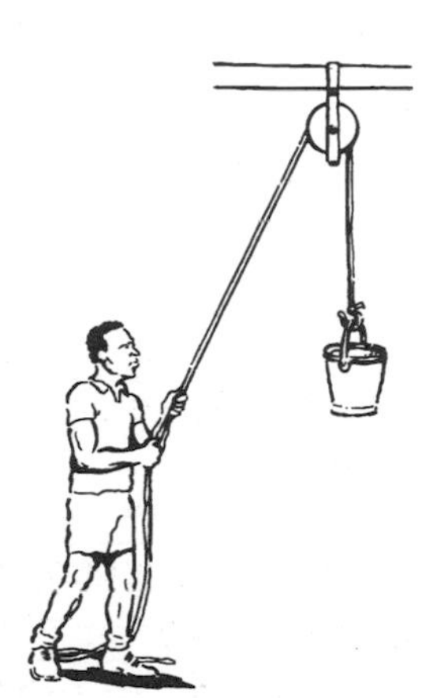

Practice questions

1) A gymnast of mass 50 kg climbs up a rope to a height of 5 m.
   a) What is the weight of the gymnast?
   b) How much work is done by the gymnast in climbing to this height?
   c) What is the gravitational potential energy of the gymnast at this height?

2) A crane raises a load at a steady speed from ground level through a vertical distance of 10 m in a time of 20 seconds. The load has a mass of 1200 kg.
   a) How much gravitational potential energy is gained by the load in 20 seconds?
   b) What is the power output of the motor of the crane during the 20 seconds of the lift?

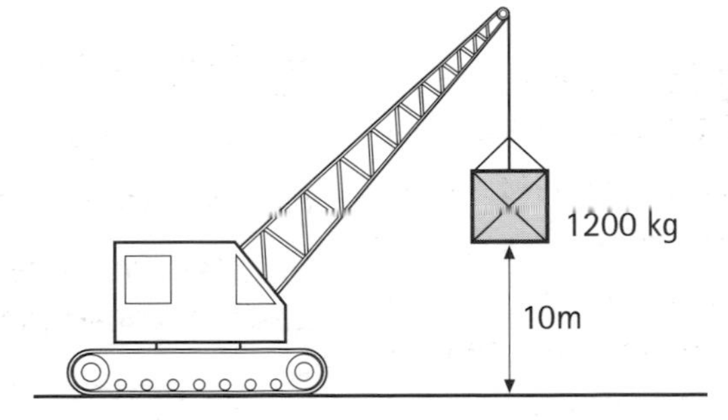

3) A cricket ball and a tennis ball are both thrown at a speed of 25 m/s. Which has the greater kinetic energy? Give a reason for your answer.

4) C A sprinter of mass 70 kg has a kinetic energy of 3.5 kJ. What is the speed of the sprinter?

Energy transfer and conservation

Energy transfer

In Standard Grade Physics you need to know about chemical energy, kinetic energy, gravitational potential energy and heat energy. You also need to be able to describe what is happening in situations where these different forms of energy are being transferred.

Can you describe what is happening to the skateboarder in this diagram in terms of different forms of energy and the transfer of energy?

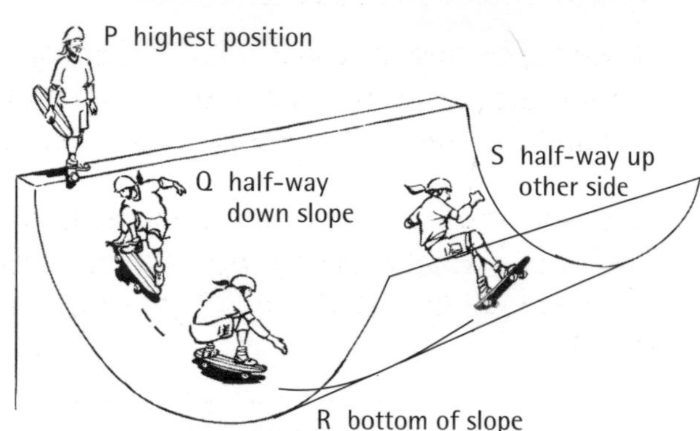

The skateboarder, Chloe, starts from rest at point P on one side of a smooth track and lets gravity take over.

She rolls down the slope through point Q and reaches the bottom of the slope at R.

She then rolls up the slope on the other side through point S.

When she rolls down the first slope, she loses gravitational potential energy and gains kinetic energy.

At the bottom of the slope, she has zero gravitational potential energy and maximum kinetic energy.

As she rolls up the second slope, there is a loss in kinetic energy and a gain in gravitational potential energy.

At her highest point on the second slope, all of Chloe's energy is in the form of gravitational potential energy.

What energy transfers take place at positions Q and S on the track?

Conservation of energy

The Law of Conservation of Energy says that you can neither create energy nor destroy it. You can transfer energy, but the total amount of energy in existence remains the same.

At all times during Chloe's motion on the skateboard the total amount of energy remains the same.

The track is perfectly smooth, so you can ignore the effect of friction and describe the motion using equations.

When Chloe is on the way down:

loss in E_p = gain in E_k

When Chloe is on the way up:

loss in E_k = gain in E_p

In practice, her final gravitational potential energy will be less than her original gravitational potential energy. This is because throughout her motion, work is being done against frictional forces and some of the energy of the system is transferred as heat energy.

However, because of conservation of energy, at any time during Chloe's motion the sum of the gravitational potential energy (E_p), the kinetic energy (E_k) and the energy that has been transferred as heat energy (E_h) is the same.

This table shows the amount of energy in the form of gravitational potential energy (E_p), kinetic energy (E_k) and heat energy (E_h) of the skateboarder system at points P, Q, R, S and T (the highest point reached on the second slope).

	E_p in joules	E_k in joules	E_h in joules
P	2500	0	0
Q	1250		150
R		2200	
S		1100	400
T	2000		500

Use the Law of Conservation of Energy to complete the table.

Practice questions

1) The skateboarder, Chloe, on the track in the diagram on page 30 (opposite) rolls from a height of 5 m. She has a mass of 50 kg. Ignore the effect of friction and air resistance on the motion.

   a) What height does Chloe reach on the other side of the track?

   b) What is Chloe's kinetic energy at point R?

   c) What is Chloe's speed at point R?

2) A man of mass 60 kg swings on a trapeze. He starts the swing from a platform 20 m above the ground.

   The lowest point reached by him during the swing is 16.8 m above ground level.

   Ignore the effect of air friction on the motion. Calculate:

   a) the man's kinetic energy at the lowest point in the swing

   b) the man's maximum speed during the swing.

Heat energy

Conduction, convection and radiation

Heat is a form of energy and is measured in joules (J). Heat is something that flows from a hot body to a cold body. Heat energy can be moved from place to place by three processes: conduction, convection and radiation.

Conduction of heat energy takes place mainly in solids.

The heat from the source causes the atoms of the solid to vibrate and gain kinetic energy. These atoms then cause neighbouring atoms to vibrate. Kinetic energy is transferred from one atom to the next. In this way, heat energy is conducted through the solid. As the atoms of the solid gain kinetic energy, the temperature of the solid increases.

REMEMBER Make sure you know the difference between temperature and heat. Temperature is a measure of something's 'hotness' or 'coldness'. It can be measured using a thermometer with a Celsius scale. Heat is a form of energy and is measured in joules.

In liquids and gases (fluids), heat energy is transferred mainly by a process called **convection**.

The heat source warms the fluid, causing it to expand and rise upwards. Cooler fluid flows in to take the place of the warm fluid. The cooler fluid is heated and it, in turn, expands and rises upwards. A circulating current of warm fluid is set up. The circulating current of warm fluid is called a 'convection current'. A roomful of air can be heated by using a convector heater as shown in the diagram.

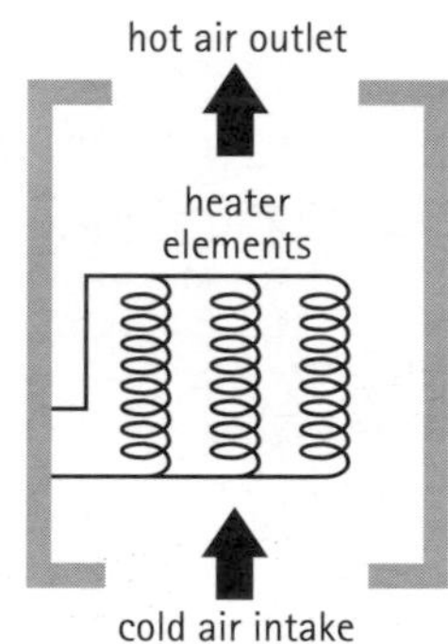

How do the convection currents produced by the heater warm up the whole room?

In **radiation**, infrared waves are emitted by the hot object at the speed of light. The infrared radiation is absorbed by a cooler body. The absorption of infrared radiation causes the temperature of the cooler body to increase. Infrared radiation is the way that heat energy reaches us from the Sun.

Loft insulation or under floor insulation reduces heat losses by conduction. The insulation material traps air. Air is a poor conductor of heat. The fibres of the insulation also prevent the trapped air from circulating and causing a convection current to be set up. Cavity wall insulation has a similar effect.

With double glazing, the insulating properties of air reduces heat losses by conduction.

Radiation losses can be reduced by using special plasterboard which is backed with a sheet of aluminium foil. The shiny metal foil reflects some of the infrared radiation back into the room and so heat losses by radiation are reduced. Metal foil placed behind a radiator which is fixed to an outside wall also helps to reduce heat losses by radiation.

Specific heat capacity and calculating heat energy

When you supply heat energy to a substance, you usually cause its temperature to increase. The change in temperature of the substance depends on the mass of the substance and the quantity of heat you supply to it. The change in temperature also depends on a property of the material called its 'specific heat capacity'.

Different amounts of heat energy are needed to raise the temperature of the same mass of different substances by 1 degree celsius. For example, 1 kilogram of water needs 4180 joules to raise its temperature by 1 degree celsius, but 1 kilogram of lead needs only 128 joules.

The specific heat capacity (c) is the heat energy that is needed to raise the temperature of a specific mass of the substance (i.e. 1 kilogram) through 1 degree celsius.

Specific heat capacity is measured in joules per kilogram per celsius degree (J/kg °C).

The heat energy (E_h) which has to be supplied to a substance of a given mass (m) and with a specific heat capacity (c) to raise its temperature by a certain amount (ΔT) is given by the equation:

$$E_h = cm\ \Delta T$$

E_h is measured in joules (J).
m is measured in kilograms (kg).
ΔT is measured in degrees celsius (°C)
c is measured in joules per kilogram per °C (J/kg °C).

You use the same equation to work out the heat which is given out by a substance as it cools.

REMEMBER You need to know how to use the relationship $E_h = cm\ \Delta T$ to calculate the heat **gained** or **lost** by a substance.

REMEMBER The temperature of boiling water is 100°C and the temperature of melting ice is 0°C.

Practice questions

1) A concrete block in a storage heater has a mass of 2 kg. The block is heated and its temperature rises from 15°C to 40°C. How much heat energy is absorbed by the block? The specific heat capacity of concrete is 850 J/kg °C.

2) An electric kettle holds 0.75 kg of water at a temperature of 15°C. The power of the heating element in the kettle is 1.5 kW. The kettle is switched on. The specific heat capacity of water is 4180 J/kg °C.

   a) How much energy is needed to raise the temperature of the water to its boiling point?

   b) Assuming that no heat energy is transferred to the surroundings, how long will it take the water to reach its boiling point?

Change of state and latent heat

Change of state

When you supply heat energy to a substance, you can cause it to change state. For example, if you supply heat to a kettle of water, you will cause the temperature of the water to rise. However, the temperature does not keep rising. At 100°C the heat energy supplied is used to make the water change from a liquid into a vapour. The energy being supplied is used to bring about a change in state. Water at 100°C is changed to steam at 100°C.

REMEMBER When a substance changes its state, the temperature of the substance does not change (e.g. boiling water at 100°C changes state to becomes steam at 100°C).

When the steam changes back to water, the heat that was supplied is given out.

Steam at 100°C has more heat energy than the same mass of water at 100°C.

When an egg is being boiled, the temperature of the water is 100°C. Will the egg cook more quickly by turning up the gas?

A change of state from liquid to vapour is called **vaporisation**. A change of state can also involve changing from a vapour to a liquid. This is called **condensation**.

Similarly, if you supply heat to a solid, you can cause the solid to change its state. For example, if you supply heat to ice at 0°C it changes to water at 0°C. This is called **fusion**. Change of state can also involve the substance changing from a liquid to a solid. This is known as **freezing**.

Latent heat

The heat energy that is taken in or given out by a substance when it changes state is called **latent heat**. When a substance changes from a solid to liquid, the latent heat involved is called the latent heat of fusion. When the substance changes from a liquid to a vapour, latent heat of vaporisation is involved.

Why is the heat involved in a change of state called latent heat?

'Latent' means hidden. The heat involved in change of state does not produce a change in temperature, so in a way its effect is concealed or hidden. A lot of heat energy is needed to change water into steam – much more than is needed to bring that same mass of water to its boiling point. This is why a burn from steam can be very severe. A large quantity of latent heat is given out when the steam condenses to form water on the skin.

Specific latent heat of fusion and vaporisation

These tell you the heat energy needed to change the state of a specific mass of a substance (i.e. 1 kilogram) from a solid to liquid and from liquid to vapour.

The **specific latent heat of vaporisation** (l) of a substance is the heat you need to change the substance from a liquid at its boiling point into vapour at the same temperature. For example, the specific latent heat of vaporisation of water is 22.6×10^5 J/kg.

The specific latent heat of fusion (l) of a substance is the heat you need to change a mass of 1 kilogram of the substance from a solid at its melting point into liquid at the same temperature. For example, the specific latent heat of fusion of ice (solid water) is 3.34×10^5 J/kg.

Which requires the greater amount of energy: the melting of 1 kg of ice or the vaporisation of 1 kg of boiling water?

Specific latent heat of fusion and specific latent heat of vaporisation both have the same symbol (l) and are measured in joules per kilogram (J/kg).

The heat energy (E_h) you need to change a mass (m) of liquid at its boiling point into vapour at the same temperature is shown as:

$$E_h = ml$$

E_h is the heat supplied in joules (J).
m is the mass in kilograms (kg) of the liquid changing state.
l is the specific latent heat of vaporisation measured in joules per kilogram (J/kg).

The same relationship ($E_h = ml$) applies to the heat energy (E_h) you need to change a mass (m) of substance in its solid state at its melting point to a liquid at the same temperature. The only thing you have to remember for this case is that the symbol l stands for the specific latent heat of fusion not specific latent heat of vaporisation.

REMEMBER At Credit Level, you need to be able to do calculations on latent heat using $E_h = ml$

Practice questions

1) How many joules of heat are produced when 0.005 kg of steam at 100°C changes into water at the same temperature?

2) How much heat energy is needed to melt an ice cube of mass 0.008 kg?

Examination questions

Try these two questions from past exam papers. Question 1 is from a **General Level** paper and question 2 is from a **Credit Level** paper. Spend about 8 minutes on question 1 and no more than 10 minutes on question 2.

When you have finished, turn to page 94 for the answers and the marking guide.

1) A heating engineer designs a heating system for a house. The engineer suggests to the householder that the radiator in the living room should be able to raise the temperature of the air in the room by 20 degrees celsius.

The mass of air in the room is 80 kilograms.

a) Calculate the energy which the radiator supplies to raise the temperature of the air by 20 degrees celsius (specific heat capacity of air = 1000 joules per kilogram per degree celsius). **(2)**

b) The living room has two outside walls. The living room also has two inside walls which back onto other heated rooms in the house as shown here.

Outside 0°C
Living room 20°C
Room 1 18°C
Room 2 18°C
Room 3 18°C

Explain why heat is transferred more quickly through the outside walls than through the inside walls of the house, even though the outside walls are thicker. **(2)**

c) The engineer says that it is important to reduce heat losses. One way of doing this is to put in wall insulation. The graphs below show how the temperature in a room falls from 20 degrees celsius, after switching off the heating, when three different types of foam X, Y and Z are used to insulate the walls.

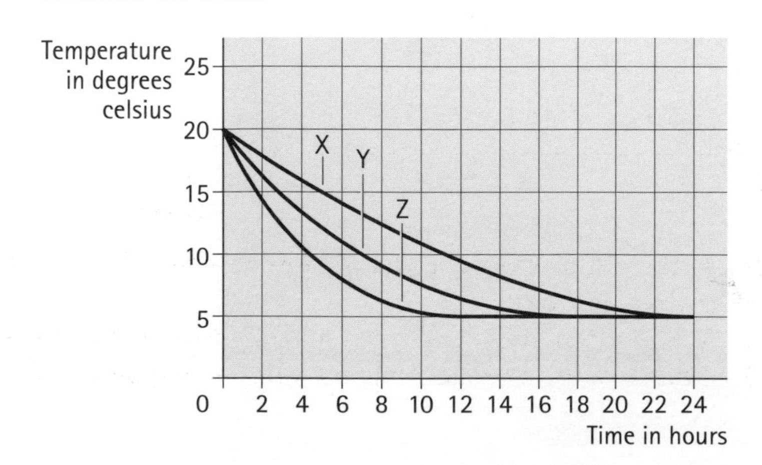

i) Explain which of the foams X, Y or Z would best reduce heat losses. **(2)**

ii) What was the outside temperature when the data for the graphs was collected? **(1)**

2) The diagram shows a water chute at a leisure pool. The top of the chute is 11.25 m above the edge of the pool. A girl, of mass 50 kg, climbs from the edge of the pool to the top of the water chute.

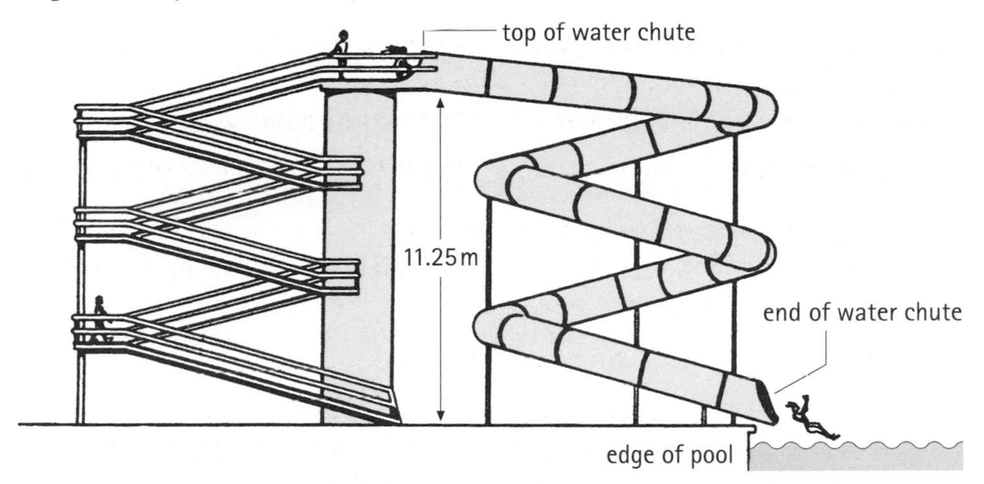

a) Calculate the potential energy gained by the girl in climbing from the edge of the pool to the top of the chute. **(2)**

b) The girl slides from rest to the bottom of the chute. Assuming that her potential energy is all transferred to kinetic energy, show that her speed at the bottom of the chute is 15 m/s. **(2)**

c) Frictional forces act on the girl so that her actual speed at the bottom of the chute is 12 m/s. The graph below shows how the girl's speed varies with time as she slides down the chute.

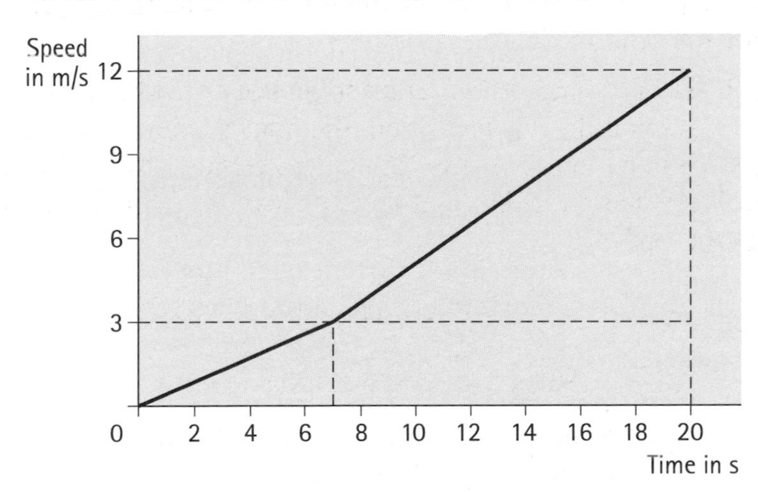

i) Calculate the distance travelled by the girl in sliding from the top to the bottom of the chute. **(3)**

ii) The energy transferred as heat in her journey down the chute is 2025 J. Calculate the average frictional force acting on the girl. **(2)**

Electricity

This section is about:

- using correctly the key words (in bold) below
- drawing circuit diagrams and using electrical symbols
- solving problems on electrical circuits
- describing how electricity is generated and transmitted.

Electricity is described in terms of **current, charge** and **voltage**. Electrical charge, carried by **electrons**, can be made to flow through **conductors**. The flow of charge, i.e. the current in the conductor, is produced by a voltage supply such as a **battery** or the **mains**. A complete path or circuit is needed before a current can be produced.

There are two types of electrical circuit: **series** and **parallel**. You can show electrical circuits using **electrical symbols** in **circuit diagrams**.

Electrical appliances draw currents of different size from the mains supply. The appliances have different **resistance**. A lower resistance path means that a larger current is produced.

High current appliances transfer **electrical energy** at a greater rate than those carrying small currents. The rate at which an appliance transfers electrical energy is called the **power** of the appliance. Different electrical appliances have different power ratings (e.g. an electric kettle is 20 times more powerful than an electric lamp).

Electricity is useful, but it can also be dangerous if it is not treated with respect. The human body is a conductor of electricity. A small current through your body can be fatal. Low voltage supplies can be handled safely, but higher voltages such as that from the mains supply can be very dangerous. An **earth wire** is connected to electrical appliances to protect you should the casing of the appliance become live. Some electrical appliances are designed so that they do not need an earth wire connection. These appliances are protected by a double layer of **insulation**.

Electricity also produces a **magnetic effect**. This is known as **electromagnetism**. This is used in **electromagnets** and **electric motors**.

Alternating current generators connected to **transformers** and power lines enable electricity to be transmitted via the **National Grid** over long distances to our homes.

FactZONE

Electrical symbols

cell

battery

fuse

lamp

switch

resistor

variable resistor

voltmeter

ammeter

Circuits

In a **series** circuit:

- the current is the same at all points
- the sum of the voltages across components is equal to the voltage of the supply.

In a **parallel** circuit:

- the sum of the current in parallel branches is equal to the current from the supply
- the voltage across components in parallel is the same for each.

Current and charge

Current is a flow of charge.

$$\text{current } (I) = \frac{\text{charge moved } (Q)}{\text{time } (t)} \qquad I = \frac{Q}{t}$$

$$Q = It$$

Charge (Q) is measured in coulombs (C).
Current (I) is measured in amperes (A).

Resistance

$$\text{resistance } (R) = \frac{\text{voltage } (V)}{\text{current } (I)} \qquad R = \frac{V}{I}$$

Resistance (R) is measured in ohms (Ω).

Resistance of resistors in series and in parallel:

(C) $R_T = R_1 + R_2$ series

(C) $\frac{1}{R_T} = \frac{1}{R_1} + \frac{1}{R_2}$ parallel

Electrical power

$$\text{power} = \frac{\text{energy transferred}}{\text{time taken}}$$

power = voltage × current $P = VI$

(C) power = current² × resistance $P = I^2R$

Transformer relationships

$$\frac{\text{secondary voltage}}{\text{primary voltage}} = \frac{\text{turns on secondary}}{\text{turns on primary}}$$

$$\frac{V_S}{V_P} = \frac{n_S}{n_P}$$

(C) $$\frac{\text{secondary current}}{\text{primary current}} = \frac{\text{turns on primary}}{\text{turns on secondary}}$$

$$\frac{I_S}{I_P} = \frac{n_S}{n_P}$$

Efficiency

(C) $$\text{efficiency} = \frac{\text{energy output}}{\text{energy input}}$$

$$= \frac{\text{power output}}{\text{power input}}$$

Kilowatt hour (kWh)

This is the 'unit' of electrical energy that electrical suppliers use to calculate your electricity bill.

Energy transferred in kWh = power of appliance (in kilowatts) × time switched on (in hours)

Earth wire and fuses

An earth wire and a fuse act as safety devices. Make sure you know how they work.

Charge, current and voltage

Charge and current

Cells, batteries and the mains supply are our common sources of electrical energy. If you connect a lamp across the terminals of a battery, electrical charge flows from the battery through the lamp. The charge transfers electrical energy to light and heat energy as it flows through the lamp. The moving charge is called a 'current'.

Current is measured in amperes (A). Charge is measured in coulombs (C).

$$\text{Current in amperes} = \frac{\text{charge moved in coulombs}}{\text{time taken in seconds}}$$

REMEMBER You should know that an electrical current is a flow of charge. The charge is carried by electrons.

A current of 1 ampere means that 1 coulomb of charge is moved every second.

The current in a lamp is 4 amperes. Complete this calculation to work out how much charge is moved through the lamp in 5 seconds.

$$\text{current} = \frac{\text{charge moved}}{\text{time}}$$

charge moved = current × time

= 4 × ....

= ☐ coulombs

The movement of charge through the lamp is continuous. Charge flows from the battery through the lamp, transferring its energy as heat and light. When the charge returns to the battery, it receives more energy. It then transfers this energy as heat and light again as it moves through the lamp and so on.

The battery and the lamp make a circuit. A current can only be produced in the lamp if this circuit is complete.

Think of this battery-lamp circuit as a ski slope. The skiers are pulled to the top of the slope. They are supplied with energy. From the top of the slope, they ski downwards, transferring their gravitational potential energy to kinetic energy. They then join the ski lift at the bottom of the slope, where they receive a fresh supply of gravitational potential energy.

The ski slope is a model of a simple circuit. Which parts of the ski-slope model represent the battery, the charge, the lamp, the circuit and the current?

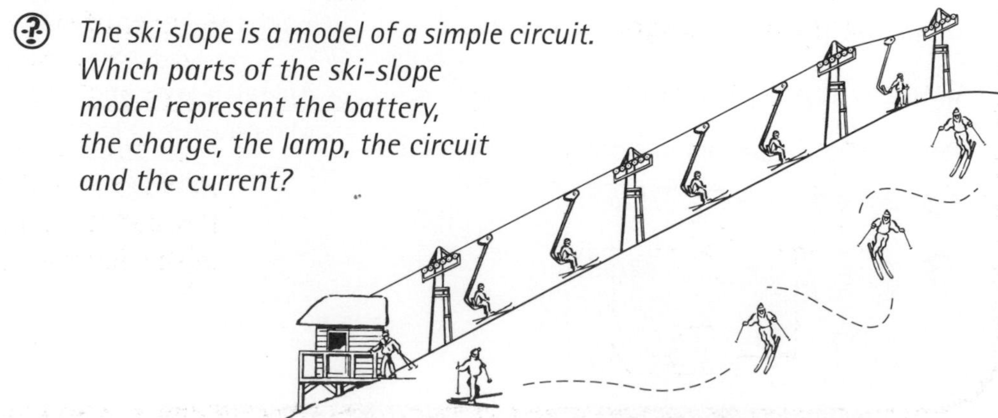

The current supplied by a battery is called direct current (d.c.). With d.c. the charges flow in one direction only.

The current from the mains supply is alternating current (a.c.). With a.c. the charges flow backwards and forwards, i.e. they alternate. In the UK, the frequency of the a.c. mains supply is 50 hertz.

Imagine that each skier takes 10 seconds to descend the slope (the lamp).

$$\text{current (skiers moved per second)} = \frac{\text{charge moved}}{\text{time}}$$

$$= \frac{1}{10}$$

$$= 0.1 \text{ coulombs per second}$$

$$= 0.1 \text{ ampere}$$

REMEMBER You can measure current with an ammeter Ammeters have to be connected in series.

Voltage

The voltage of the battery is a measure of the energy given to the charge as it moves through the battery. So, if the battery has a voltage of 12 volts, then 12 joules of energy are given to a coulomb of charge as it moves through the battery.

In the ski-slope model on page 40 (opposite), each skier could represent 1 coulomb of charge. The ski tow (the battery) would supply 12 joules of energy to each skier.

Voltage can also be a measure of the energy transferred by the charges as they move round the circuit.

Battery voltages are usually in the range of 2 to 12 volts. Batteries are made up of a number of cells connected together (+ to –) one after the other (in series). Cells have voltages of 1 or 2 volts. The voltage of the UK mains supply is 230 V.

REMEMBER You can measure voltage with a voltmeter. This is connected across (in parallel with) the component whose voltage you want to measure.

Practice question

a) Complete the symbols for the ammeter and the voltmeter for the circuit shown in the diagram.

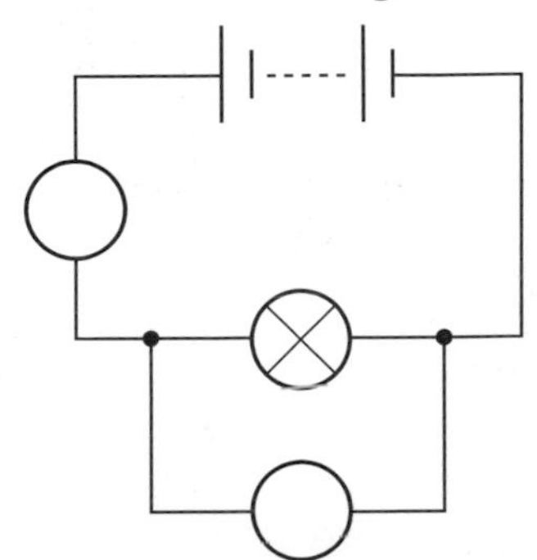

b) When a charge of 2 coulombs passes through the lamp, it transfers 6 joules of energy. What is the voltage across the lamp (shown in the diagram for part a)?

c) The current in the lamp is 2 amperes. How many coulombs of charge pass through the lamp in 3 seconds (shown in the diagram for part a)?

Series and parallel circuits

There are two basic types of electrical circuit. One is called a **series circuit** and the other is a **parallel circuit**. You can use circuit diagrams to show these two types of circuit.

Series circuit

This is an example of a simple series circuit. The components in the circuit (the battery, the two lamps and the two ammeters) are connected one after the other, i.e. in series. In a series circuit, there is only one path for the current.

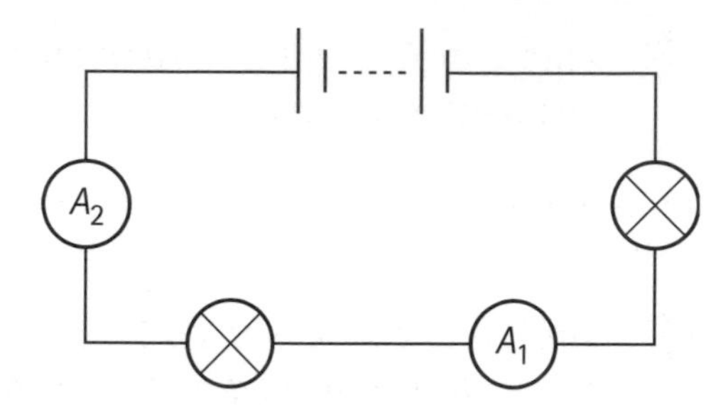

There are two important rules for a series circuit.

Rule 1 The current has the same value at all points in the circuit.

Ammeter A_1 reads 2 A. What does ammeter A_2 read?

Rule 2 The sum of the voltages across each of the components is equal to the voltage of the supply.

Voltmeter V_1 reads 4 V. The voltmeter across the battery reads 6 V. What does voltmeter V_2 read?

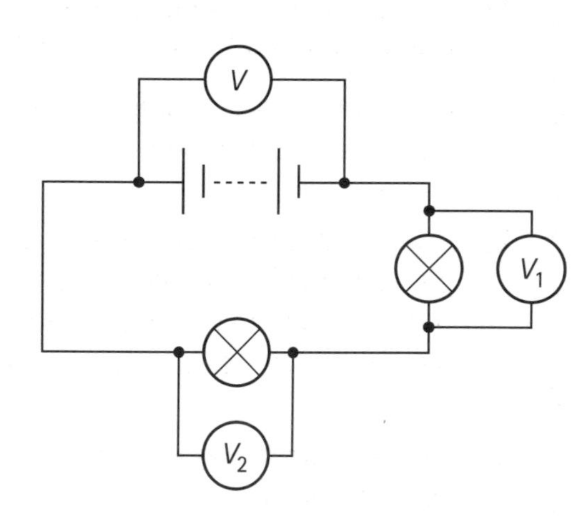

Parallel circuit

This is an example of a parallel circuit. It is different from a series circuit because there is more than one path for the current. The circuit has branches. The branches are in parallel with each other.

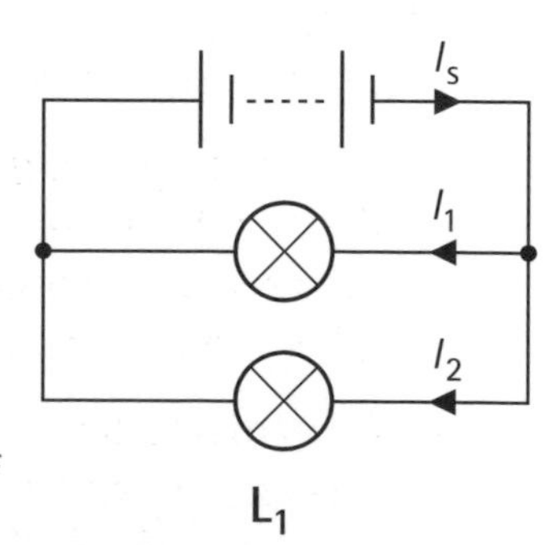

There are two important rules for a parallel circuit.

Rule 1 The current from the supply is the sum of the currents in the parallel branches.

$$I_S = I_1 + I_2$$

How many parallel branches are there in this circuit?

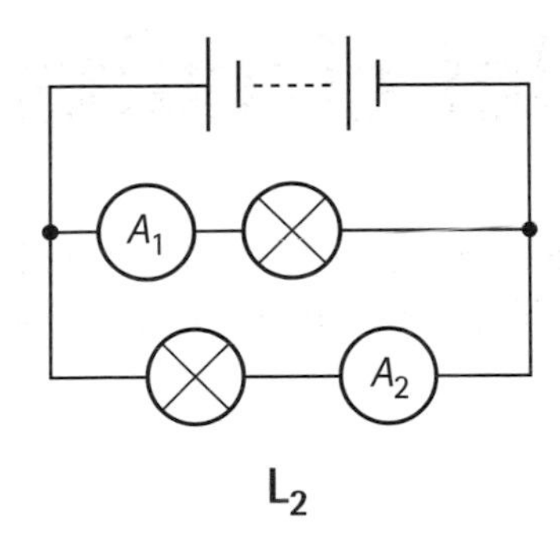

Ammeter A_1 reads 2 A. Ammeter A_2 reads 1 A. What size of current is supplied by the battery?

Rule 2 Where components are connected in parallel with each other, the voltage across each component is the same.

The voltage across lamp L_1 is 6 V. What is the voltage across L_2? What is the voltage provided by the battery?

Practice questions

1) A torch is an example of a series circuit. The circuit is made up of a battery, a switch and a lamp.

   Draw the circuit diagram for the torch circuit. Use the correct symbols.

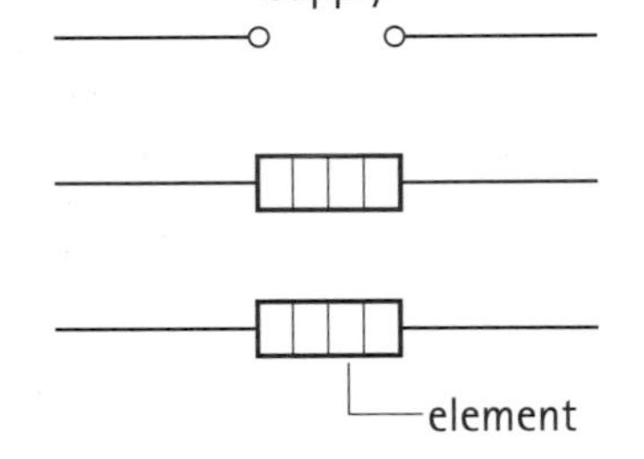

2) A two-bar electric fire is an example of a simple parallel circuit. Complete the circuit diagram for a two-bar electric fire so that either two bars or one bar can be switched on using two switches.

3) The diagram shows the rear lamps, side lamps and head lamps in a car lighting circuit.

   a) What type of circuit is represented by the diagram?

   b) Which lamps are on when:
      i) only S_1 is closed
      ii) only S_2 is closed
      iii) S_1 and S_2 are closed?

   c) A headlamp takes a current of 3 A. A side lamp and a rear lamp each take a current of 1 A. What is the current supplied by the battery when all of the lamps are on?

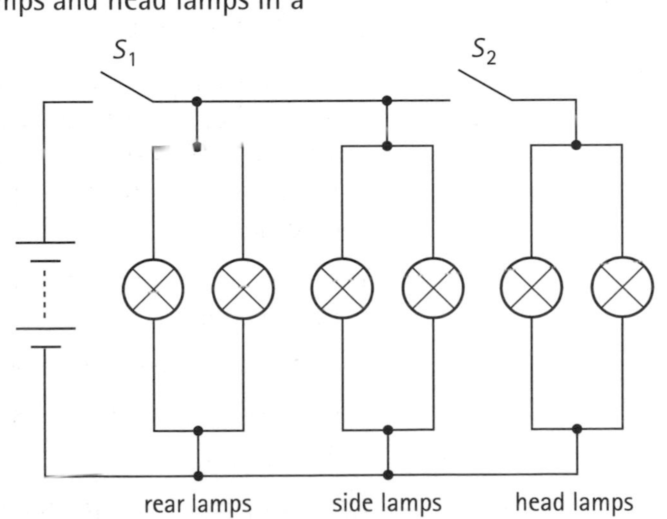

Ohm's Law and resistance

Ohm's Law

Ohm's Law is a simple rule concerning current and voltage. When you connect a piece of conducting material across a voltage supply, a current is produced in the material. The size of the current produced in the conductor depends on the size of the voltage across the conductor. In fact, the two quantities are directly proportional to each other. If you double the voltage, you double the current. If you halve the voltage, you halve the current. The result is called Ohm's Law.

This doesn't work with all conductors – but is true for metals and carbon. Ohm's Law only works if the temperature of the conductor is kept fixed.

This graph shows how the current varies with voltage for two different conductors.

Note that for any given voltage the current in conductor 1 is always less than the current in conductor 2. Test this for different values of voltage. Use the values from the graph to complete the table.

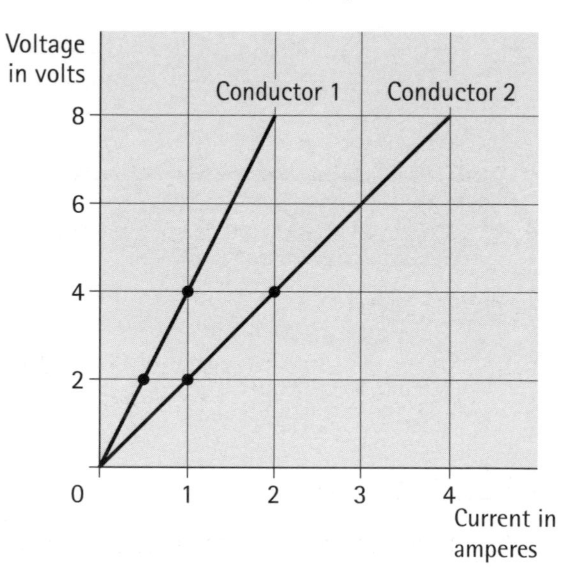

Voltage across conductors in volts	0	2	4	6	8
Current in conductor 1 in amperes	0	0.5	1		
Current in conductor 2 in amperes	0	1	2		

Resistance

Conductor 1 resists the current more than conductor 2. Conductor 1 has more resistance than conductor 2.

Resistance is the ratio of voltage to current. It is measured in ohms (Ω).

$$\text{resistance } (R) = \frac{\text{voltage across conductor } (V)}{\text{current in conductor } (I)}$$

$$R = \frac{V}{I}$$

To calculate the resistance of conductor 1, measure the voltage across the conductor and current in the conductor. You can choose any pair values you like (other than zero) for voltage and current. All pairs of values will give the same answer.

$$\text{resistance of conductor 1} = \frac{V}{R_T}$$

$$= \frac{2}{0.5}$$

$$= 4 \text{ ohms}$$

Why do all pairs of values for voltage and current give the same answer for the resistance of the conductor? (Look back to the graph of voltage against current for conductor 1. The graph has straight lines through the origin.)

Complete this calculation to work out the resistance of conductor 2.

$$\text{resistance of conductor 2} = \frac{V}{I}$$

$$= \frac{\quad}{\quad}$$

$$= 2\ \Omega$$

This is a variable resistor. Draw its electrical symbol.

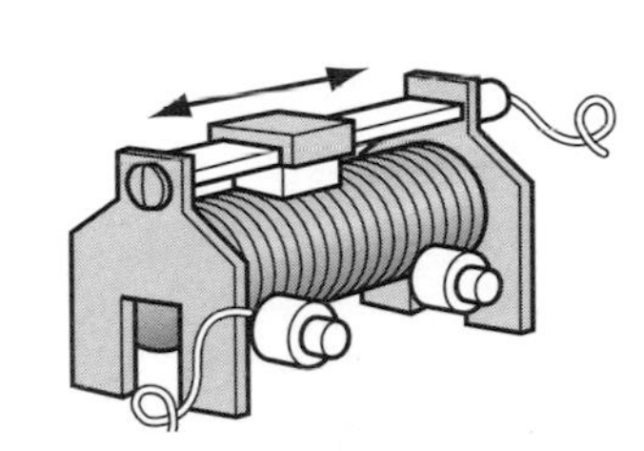

Resistors in series and in parallel

Resistors are electrical components that have resistance.

Series resistors

When two resistors of resistance R_1 and R_2 are connected in series, the total resistance R_T is given by: $R_T = R_1 + R_2$

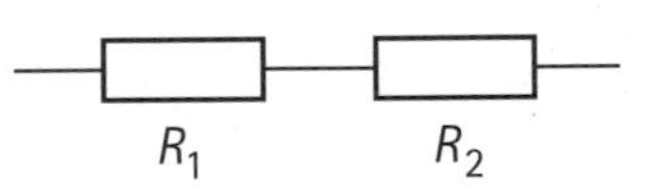

Parallel resistors

When two resistors of resistance R_1 and R_2 are connected in parallel, the total resistance R_T of the parallel combination is given by:

$$\frac{1}{R_T} = \frac{1}{R_1} + \frac{1}{R_2}$$

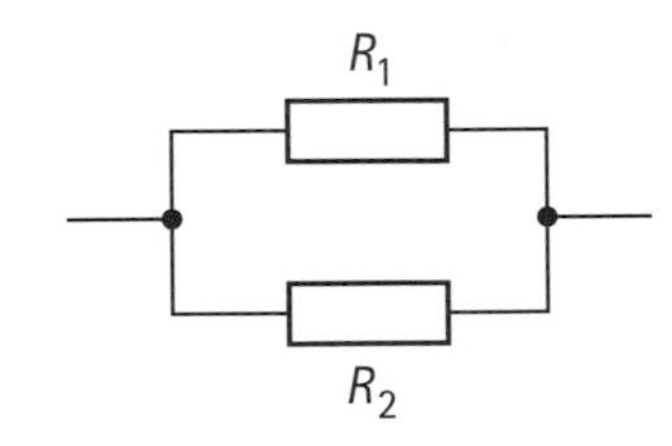

Practice questions

1) A 6 ohm resistor carries a current of 0.5 A. What is the voltage across the resistor?

2) a) Which of the circuits below are series circuits and which are parallel?

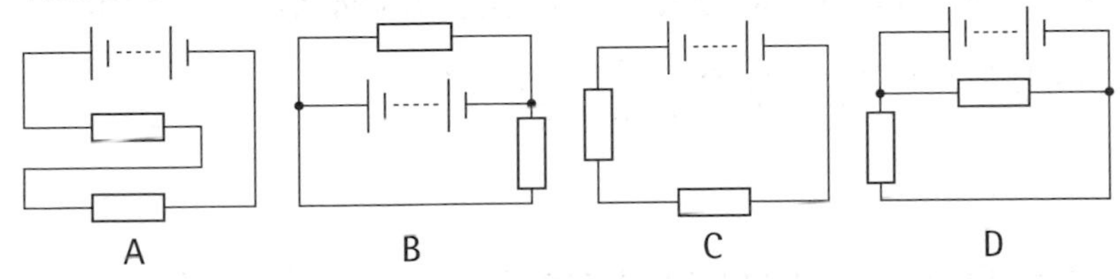

b) Each of the resistors in the circuit diagrams has a resistance of 5 ohms. What is the resistance of circuit A? What is the resistance of circuit B?

Electrical power

Energy is transferred when there is an electrical current in a circuit. When you plug an electrical appliance into the mains socket and switch it on, energy is transferred as heat, light, sound or movement. For example, a washing machine is designed to produce heat and movement. An electric lamp transfers energy from the main supply. Its purpose is to produce light, but it also produces heat. A three-bar electric fire is designed to transfer energy from the mains supply in the form of heat.

REMEMBER Typical power ratings of some electrical appliances:

electric blanket	100 W
kettle	2000 W
TV	120 W
toaster	900 W

Different electrical appliances transfer energy at different rates. Some are more 'powerful' than others. The power of an appliance is the rate at which it transfers energy from the main supply.

$$\text{power } (P) = \frac{\text{energy transferred } (E)}{\text{time taken } (t)}$$

$$P = \frac{E}{t}$$

Power is measured in joules per second or watts (W).

Power, voltage and current

The energy transferred every second by an appliance or a component in a circuit can be calculated by multiplying the voltage across the appliance by the current in it.

energy transferred per second = voltage × current

power = voltage × current

$$P = VI$$

A car vacuum cleaner that draws a current of 2 A from a supply of 12 V has a power of 24 W.

The power of the vacuum cleaner is calculated using $P = VI$

power = voltage × current = 12 × 2 = 24 W

REMEMBER At Credit Level, you can combine the equations $P = VI$ and $V = IR$ to get another expression for power. The power (P) developed by an electrical component of resistance (R) carrying a current (I) is given by:

$P = I^2 R$

Here is the algebra:

$P = VI = (IR) I = I^2 R$

An electric toaster that operates from a 230 V mains supply has a power of 920 W. Complete this calculation to find the current drawn by the toaster.

$P = VI$

$I = \frac{}{}$

$I = \frac{920}{}$

= ☐ amperes

Fuses and household circuits

A fuse protects the flex that connects the appliance to the mains socket. If there is a fault and too much current is drawn by the appliance, there is a danger that the flex will overheat, causing a fire. The plug of the appliance has a fuse fitted. If the current drawn gets too big, the fuse melts and the circuit is broken. The rating of the fuse in the plug is matched to the power rating of the appliance.

Fuses also protect the circuits in our homes. Very high power appliances, such as electric cookers and electric showers, are connected to the mains supply by a separate special circuit. The circuit has thick wiring so that it can carry high currents without overheating the wires. These circuits usually have 45 A fuses.

Normal mains appliances use a special kind of parallel circuit called a 'ring circuit'. The ring circuit is protected by a 30 A fuse.

Sometimes, circuit breakers are used for household circuits instead of fuses. Circuit breakers act like automatic switches. They sense when the current is too large and cause a break in the live wire.

C The ring main circuit uses cable that has three wires: live (L), neutral (N) and earth (E). The wires are connected to the terminals in the mains sockets as shown in the diagram.

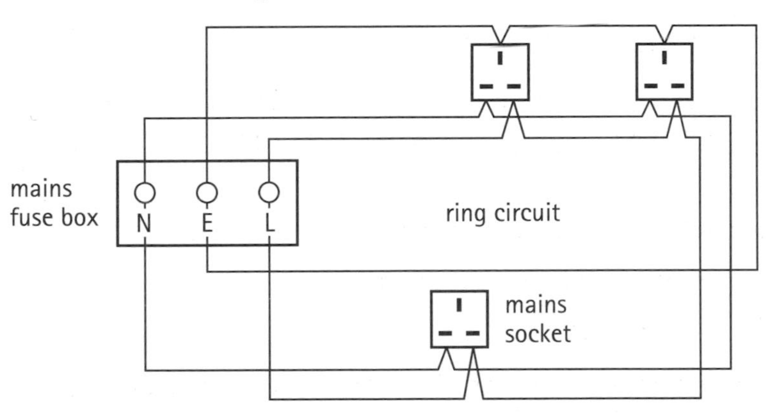

Each wire forms a ring which starts and finishes at the mains fuse box. The ring circuit allows the current to be delivered by two routes to the appliances connected to the mains socket. So the current is shared by two wires. The advantage of this design is that thinner (so less costly) wiring, can be used to carry the current safely without danger of overheating in the wiring.

? *The lighting circuits in your home use a normal parallel circuit protected by a 5 A fuse. Why does the household lighting circuit have a fuse of a smaller rating than the ring main circuit?*

REMEMBER Appliances up to 700 W use 3 A fuses in their plugs. Higher power appliances have 13 A fuses fitted.

REMEMBER Some electrical appliances have an earth wire connection, others are protected by a double layer of insulation. They are marked with this double insulation symbol.

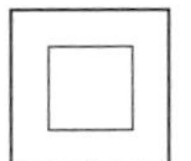

REMEMBER Sometimes, circuit breakers are preferable to fuses because they operate more quickly and only need to be reset rather than replaced.

Practice question

An electric kettle is rated at 230 V, 1.5 kW.

a) What is the current in the kettle when it is switched on?

b) What size of fuse is fitted in its plug?

c) What is the purpose of the earth wire in the plug?

Electromagnetism

Magnetism from electricity

A current in a wire produces a magnetic effect. If the wire is wound into a coil, the magnetic effect becomes stronger. The magnetic field of the current-carrying coil is like the field of a bar magnet. A current-carrying coil is called a 'solenoid' or 'electromagnet'.

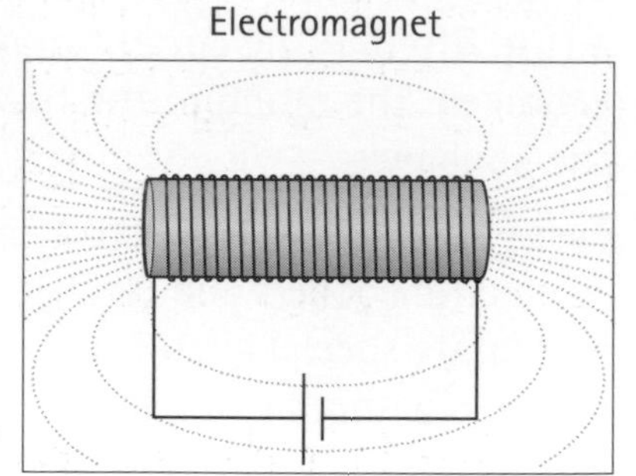

> **REMEMBER** Electromagnetism is the name used to describe the relationship between electricity and magnetism.

The magnetic field of a current-carrying coil of wire can be made to act with the magnetic field of a magnet to produce a magnetic force. The magnetic forces on a current-carrying coil in a magnetic field can be used to make the coil rotate in the magnetic field. This is the principle of the **electric motor**.

This diagram shows the four important parts of an electric motor – coil, commutator, brushes and magnet. Label the parts on the diagram.

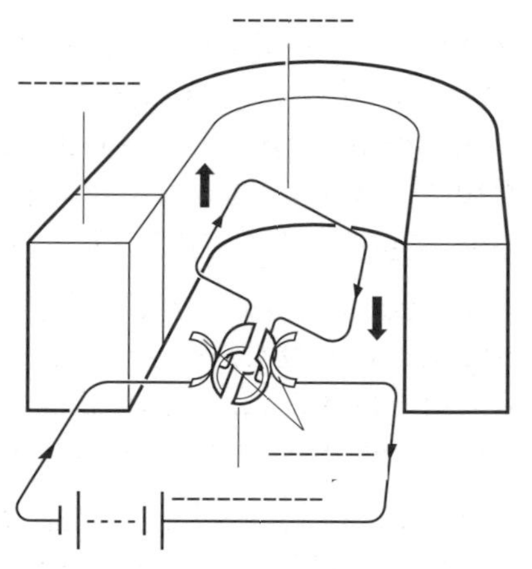

> **REMEMBER** At Credit Level, you are expected to be able to describe how these parts enable the motor to work.

When the coil is horizontal (diagram A, below), the commutator makes contact with the brushes and a current is supplied to the coil. The magnetic field of the magnet and the magnetic field due to the current in the coil cause magnetic forces to act on the coil. An upward force acts on one side of the coil and a downward force acts on the other side of the coil.

The combination of forces acting on the coil makes it rotate into a vertical position (diagram B). At this point, the commutator loses contact with the brushes and there is no current in the coil. The coil can rotate freely beyond the vertical position.

The commutator makes contact with the brushes again (diagram C). Again, a current is supplied to the coil. The combination of forces act on the coil in the same direction as before. The coil rotates to the vertical position. Again, for a moment, contact is lost. The coil continues to rotate, contact is made again with the brushes, current is then supplied to the coil, forces act and so on.

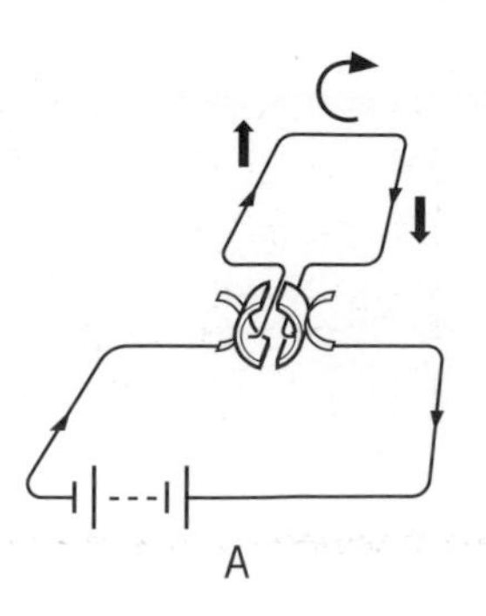

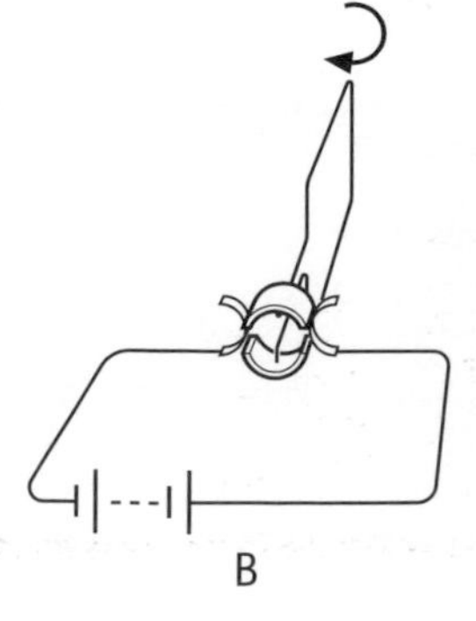

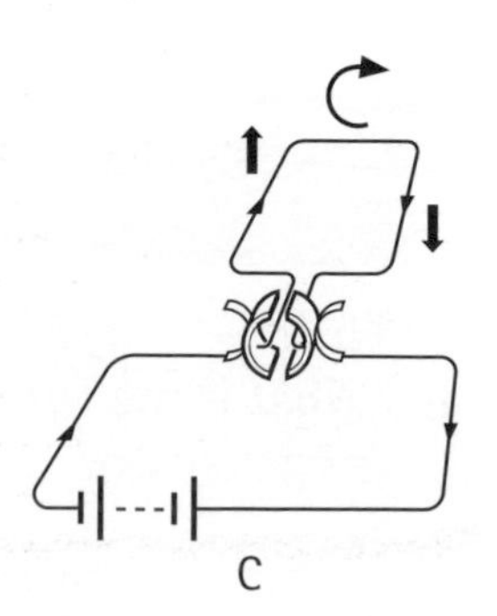

In a commercial motor, the commutator is made up of many pairs of segments. Coils with many windings are connected across each pair. This arrangement allows for a smoother rotation. The magnetic field is provided by coils carrying an electrical current (called field coils) rather than a permanent magnet. The use of field coils enables the motor to be operated using alternating current.

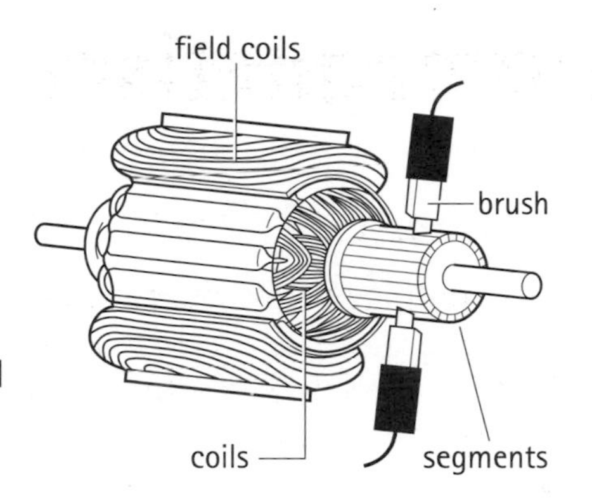

REMEMBER At Credit Level you need to know how the parts of the generator enable it to produce alternating current.

Electricity from magnetism

When a piece of wire is moved so that it cuts through a magnetic field, a voltage is induced in the wire. The size of the induced voltage is increased if the wire is wound into a coil. The size of the voltage also increases if you increase the strength of the magnetic field and the speed of the coil in the magnetic field.

A voltage is also induced when the wire or coil is held stationary and the magnetic field is moved. It is the relative motion between the coil and magnetic field that causes a voltage to be induced. This is the principle of the alternating current generator.

Generators are used in power stations to produce electricity for the National Grid. In the generator a rotating electromagnet (rotor) induces an alternating voltage in a stationary coil (stator). The alternating voltage produces an alternating current.

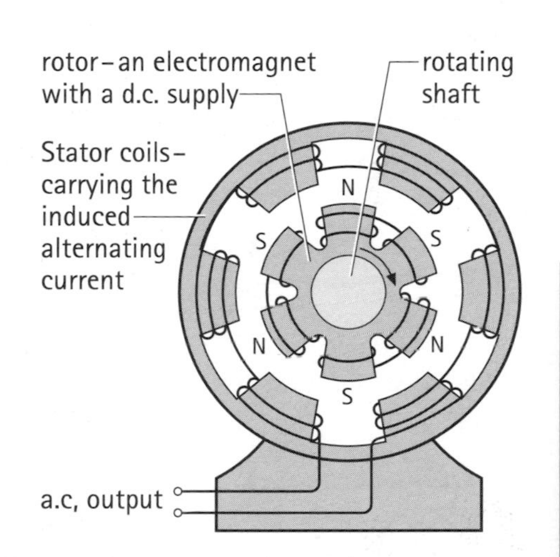

Practice question

This diagram shows how a bicycle dynamo is operated and how it is constructed.

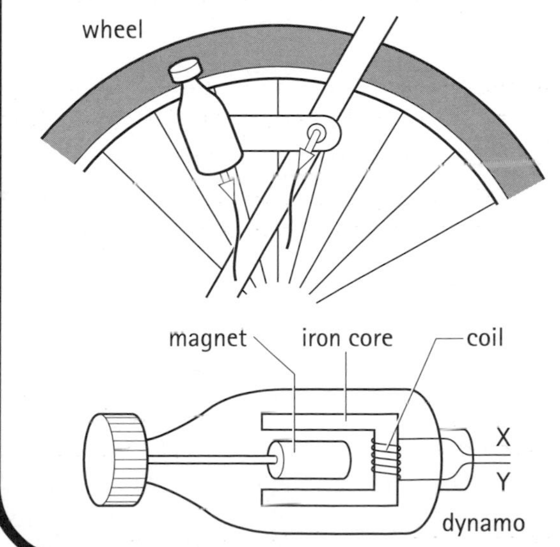

a) Describe how a voltage is produced across XY.

b) What type of current is produced by the dynamo?

c) This graph shows how the current from the dynamo varies with time when the bicycle is travelling at a steady speed.

Sketch the graph that would be obtained if the bicycle moves with a slower steady speed.

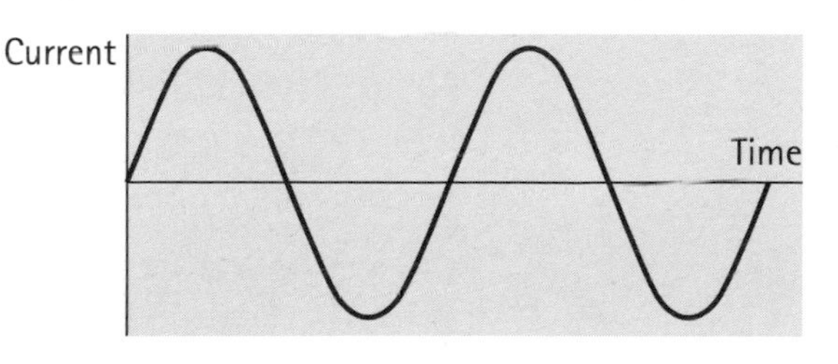

Power generation and transmission

Power stations

Power stations generate electrical energy by burning fossil fuels or by the use of nuclear fuel. The energy from the fuel is used to produce heat. The heat produces steam to drive turbines which, in turn, drive the generators that produce electricity.

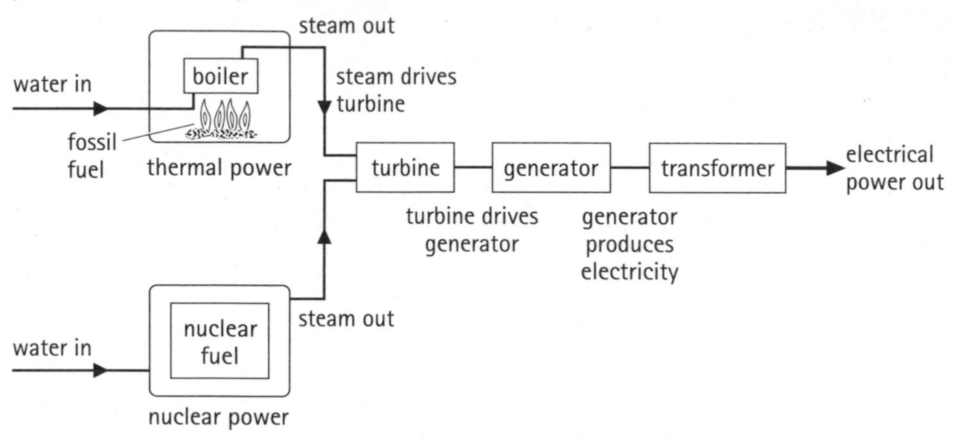

REMEMBER Efficiency has no unit. It is a fraction and is normally expressed as a percentage.

The **efficiency of a power station** is the energy transferred at its output divided by the energy it obtains from its fuel source.

$$\text{efficiency} = \frac{\text{energy transferred at output}}{\text{energy delivered at input}}$$

The energy transferred by the power station is always less than the energy delivered by the fuel. In a typical power station, only about 35% of the energy from the fuel actually reaches the consumer. This is because energy in the form of heat is transferred and lost to the surroundings at the different stages of the electricity generation process.

Transmission lines

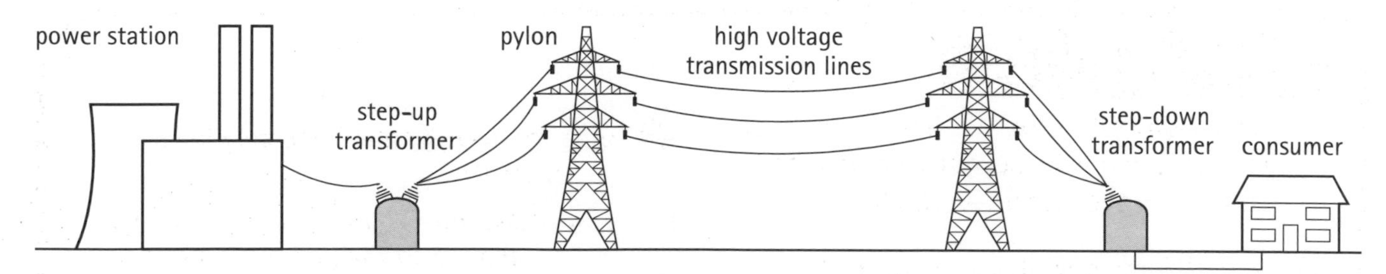

Electricity is transmitted throughout the UK via a network of transmission lines. Pylons are used to support the cables of the transmission lines. This network of pylons and transmission lines is called the National Grid.

Devices called step-up transformers link the power stations with the National Grid. The The high voltage from the transformer is applied to the electricity transmission cables. High voltages are used in order to reduce power losses in the transmission cables. Heat is produced in the cables during the transmission of electricity. Much more heat would be produced if lower voltages were used for transmission.

Step-down transformers are used at the consumer end of the transmission lines. These transformers step down the high voltage to a level that is suitable for factory machinery and the electrical appliances in your home.

> **REMEMBER** Note that step-up transformers convert (or transform) low voltages to higher voltages. Step-down transformers convert high voltages to lower voltages.

Transformers

A transformer is made of two separate coils of wire which are wound on an iron core. One is called the primary coil and the other is called the secondary coil. When an alternating voltage is applied to the primary coil, it changes the magnetic field close to the coil. This changing magnetic field induces a voltage in the secondary coil.

The ratio of the voltage V_S produced across the secondary coil and the voltage V_P supplied to the primary coil is equal to the ratio of number of turns n_S on the secondary coil and the number of turns n_P on the primary coil.

$$\frac{V_S}{V_P} = \frac{n_S}{n_P}$$

The efficiency of a transformer is the power obtained from the secondary coil divided by the power supplied to the primary coil:

$$\text{efficiency of a transformer} = \frac{\text{power obtained from secondary}}{\text{power supplied to primary}}$$

$$= \frac{V_S \times I_S}{V_P \times I_P}$$

> **REMEMBER** At Credit Level, you need to know how to carry out calculations involving efficiency and primary and secondary currents and voltages.

In a perfect transformer (i.e. one which is 100% efficient), the power obtained from the secondary coil is equal to the power supplied to the primary coil:

$$V_S \times I_S = V_P \times I_P$$

In practice, transformers are not 100% efficient. Some power losses take place due to heating in the coils and the core of the transformer.

Practice questions

1) A step-down transformer has a turns ratio of 10:1. What is its output voltage if the voltage on the primary is 230 V?

2) A 24 W lamp operates at its normal brightness when connected across the secondary of a transformer. The primary voltage is 2 V.

   (C) a) Assuming that the transformer is 100% efficient, what is the current in the primary coil?

   (C) b) If the transformer were only 96% efficient:

   i) how much power would have to be supplied to the primary coil?

   ii) what would be the current in the primary coil to make the lamp operate normally?

3) Explain why a transformer cannot be used to step down the voltage of a car battery.

Examination questions

Try these three questions from past exam papers. Questions 1 and 2 are from **General Level** papers and question 3 is from a **Credit Level** paper. Spend about 12 minutes in total on questions 1 and 2 and no more than 9 minutes on question 3.

When you have finished, turn to page 95 for the answers and the marking guide.

1) The diagram below shows the connections between a 12 volt car battery and a rear window heater.

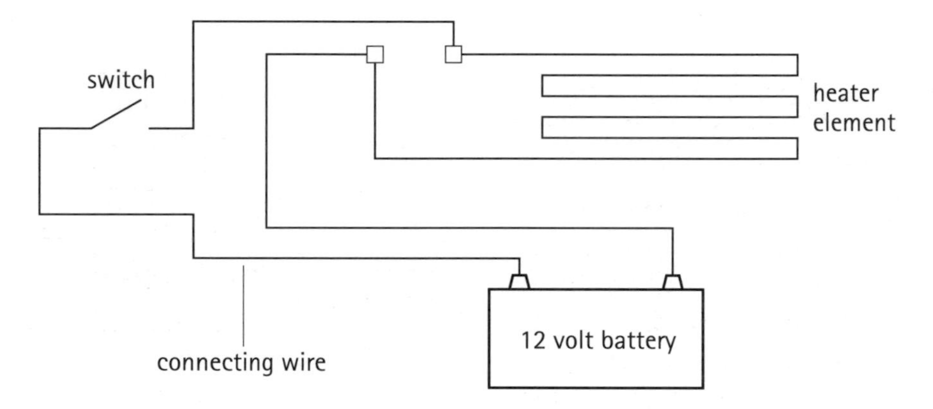

When the switch is closed, there is a current of 10 amperes in the circuit.

a) Calculate the power of the heater. **(2)**

b) What is the resistance of the heater? **(2)**

c) The heater element becomes hot, but the connecting wires remain cold. Complete the following statement by putting the phrase *less than* **or** *the same as* **or** *greater than* in the space.

The resistance of the connecting wires is .... the resistance of the heater element. **(1)**

2) a) The main sources of energy used in the UK are listed below.

Oil   Coal   Natural gas   Nuclear fuel   Hydroelectricity

i) Select from the list one source of energy which is **not** a fossil fuel. **(1)**

ii) Name a renewable source of energy which is **not** in the list. **(1)**

iii) Give an advantage of generating electricity using hydroelectric power. **(1)**

b) A power station produces electricity at 25 000 volts. This voltage is stepped up to 400 000 volts by a transformer.

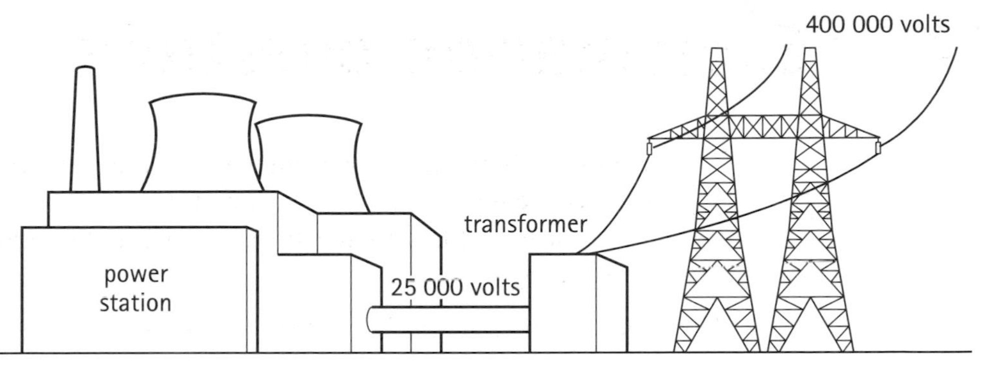

i) The number of turns on the primary coil of the transformer is 20 000. Calculate the number of turns on the secondary coil. **(2)**

ii) Why is a voltage as high as 400 000 volts used in the transmission of electrical energy? **(1)**

3) a) The diagram below shows a simple hand-operated generator which is used to light a lamp.

iron core
lamp
N
S
handle
coil
magnet

i) Explain why a voltage is induced across the coil of the generator when the handle is turned. **(1)**

ii) The induced voltage increases when the handle is rotated faster.

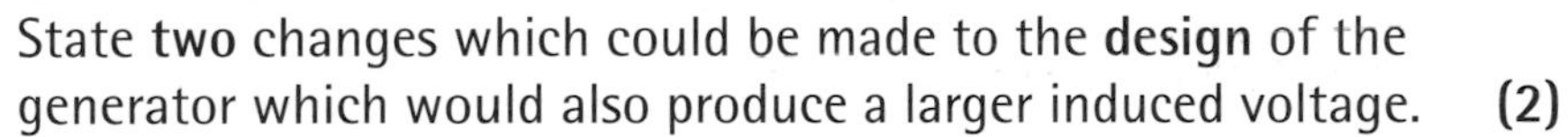

State **two** changes which could be made to the **design** of the generator which would also produce a larger induced voltage. **(2)**

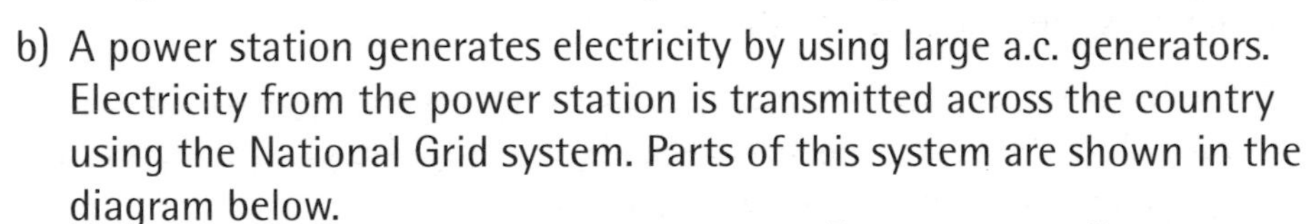

b) A power station generates electricity by using large a.c. generators. Electricity from the power station is transmitted across the country using the National Grid system. Parts of this system are shown in the diagram below.

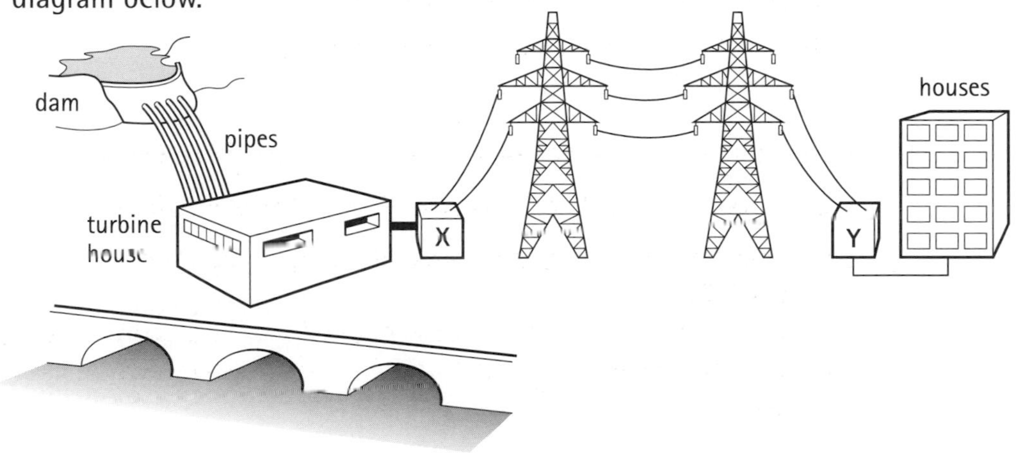

i) Name the parts labelled X and Y and describe the purpose of each. **(2)**

ii) A power line in the system has a resistance of 2 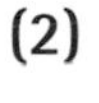Ω for every kilometre length. The power line is 100 km long and carries a current of 200 A. Calculate the electrical power loss in the line. **(3)**

Electronics

This section is about:

- using correctly the key words (in bold) below
- explaining how transistors, logic gates, clocks and counters are used in electronic systems
- calculating the resistance of components in electronic systems.

An **electronic system** is a piece of equipment that uses the movement of **electrons** to enable it to work. Video recorders, television sets, radios, mobile phones, computers and security alarms are all electronic systems.

Electronic systems are made of many different types of **components**: **resistors, capacitors, transistors, light emitting diodes, light dependent resistors, switches** and **logic gates**. Electronic systems have three basic sections: an **input** section, a **processing** section and an **output** section.

The input section is where an electrical signal is produced. An **input device** is used to produce the signal for the system. **Microphones, thermistors** and light dependent resistors are examples of input devices. **Output devices** convert the signal in the electronic system to light, sound or movement. **Loudspeakers, lamps,** light emitting diodes and **buzzers** are examples of output devices.

Some output devices are designed to cope with **analogue signals** and others are equipped to deal with **digital signals.** Digital signals can be represented by two states: **high** and **low.** An analogue signal has any number of states. Loudspeakers are output devices which are designed to deal with analogue signals. A buzzer, on the other hand, is designed to handle digital signals.

The processing section of the system 'processes' the signal from the input so that it is converted into a form the output section can deal with. An **amplifier** is an example of the processing section. The function of the amplifier is to boost the signal from the input part of the system.

Transistors are found in almost every type of electronic system. They can be used as fast-acting electronic switches.

Combinations of transistors can be connected to form logic gates. Logic gates are electronic devices which make decisions. Combinations of logic gates can be used to **process** a number of input signals so that the electronic system responds to the signals in a particular way. In an alarm system you might want a siren to sound when the alarm is switched on **and** when light **or** movement is detected by the input section of the system. A combination of an **AND gate** and an **OR gate** could be used to make the system work as you want.

Logic gates are also used in **electronic clocks** and **counting circuits.**

Electrical symbols

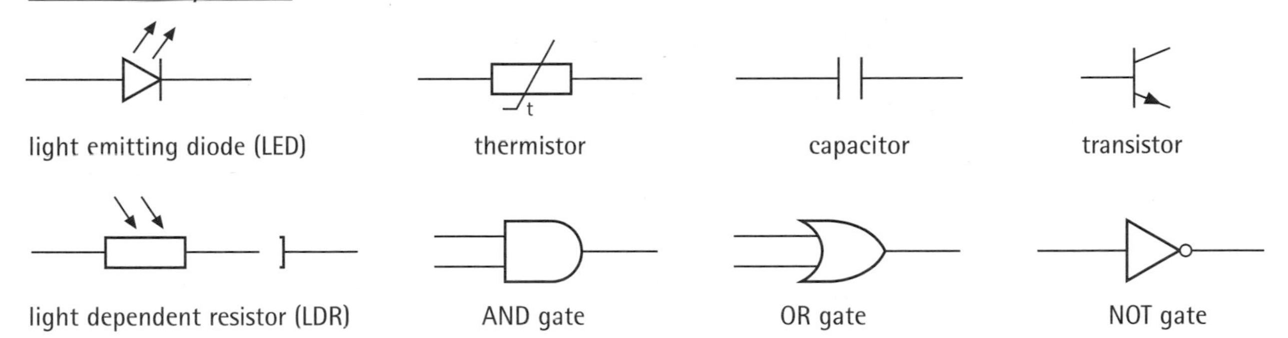

All electronic systems have a **processing section** and:

- an **input section**, which provides the input signal for the electronic system (e.g. capacitor, LDR, microphone, solar cell, switch, thermistor, thermocouple)
- an **output section**, which converts electrical signals into light or sound or movement (e.g. LED, seven segment display, loudspeaker, electric motor, solenoid, relay).

Analogue and digital signals

Analogue output changes continuously and can have any value in a given range.

Digital outputs can be represented by a high (logic 1) or a low (logic 0).

Amplifiers

An audio amplifier in a radio or TV is designed so that its output signal is a larger copy of the input signal. The output signal from an audio amplifier has the same frequency as the input signal but the amplitude is larger.

$$\text{voltage gain of amplifier} = \frac{\text{output voltage}}{\text{input voltage}}$$

$$\text{power gain of amplifier} = \frac{\text{output power}}{\text{input power}}$$

$$\text{power delivered from the output an amplifier} = \frac{V^2}{R}$$

V is the voltage across the output device (loudspeaker).

R is the resistance of the device.

Logic gates

Learn these truth tables for AND, OR and NOT gates.

INPUT		OUTPUT
A	B	Z
0	0	0
0	1	0
1	0	0
1	1	1

AND

INPUT		OUTPUT
A	B	Z
0	0	0
0	1	1
1	0	1
1	1	1

OR

INPUT	OUTPUT
A	Z
0	1
1	0

NOT

Electronic systems

Input, processing and output

Any electronic system can be shown by a block diagram of three linked parts: an input part, a processing part and an output part.

The sound equipment for a band is an example of an electronic system.

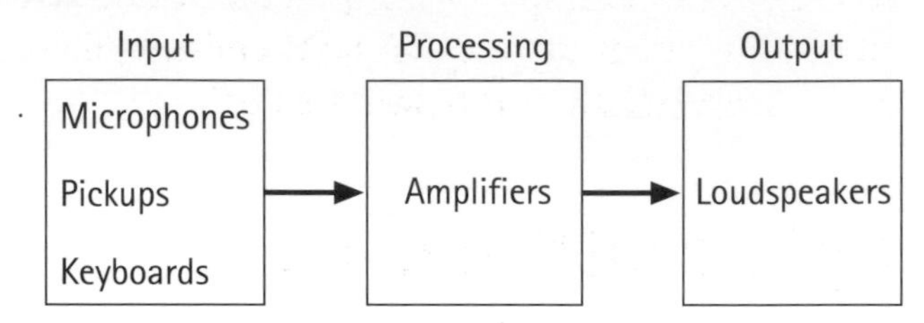

The **input** part of the sound system consists of the microphones, the guitar pickups and the keyboards.

This part is responsible for producing the electrical signals which enable the system to work. For example, the microphone converts the sound energy from the vocalist into electrical signals. The guitar pickups produce electrical signals representing the musical notes made by the guitars. Electrical signals representing musical notes are also produced by the keyboards.

REMEMBER The ratio of the output voltage from an amplifier to the input voltage is known as the 'voltage gain of the amplifier'.

$$\text{Voltage gain} = \frac{\text{output voltage}}{\text{input voltage}}$$

The **processing** part of the system is the amplifier. In this part, the signals from the input part are changed or processed so that they are capable of operating the output part of the system. The amplifier processes the input signals by increasing their voltage to a level that is suitable for the output part of the system.

The **output** part of the system converts the processed electrical signals into the form that suits the purpose of the system. The purpose of the band's sound equipment is to amplify the music for the audience. The output part of the system uses the processed signals to send out amplified sound. The output part of the band's sound system is the loudspeakers.

Digital and analogue signals

Electronic systems can process signals that are in either a digital form or an analogue form.

A digital signal gives this pattern on the screen of a cathode ray oscilloscope (CRO). Digital signals can be represented by two states.

high | on
low | off

The signals can be 'on' or 'off' or they can be 'high' or 'low'. Digital signals are like steps and have no in-between values.

An analogue signal has a pattern like this.

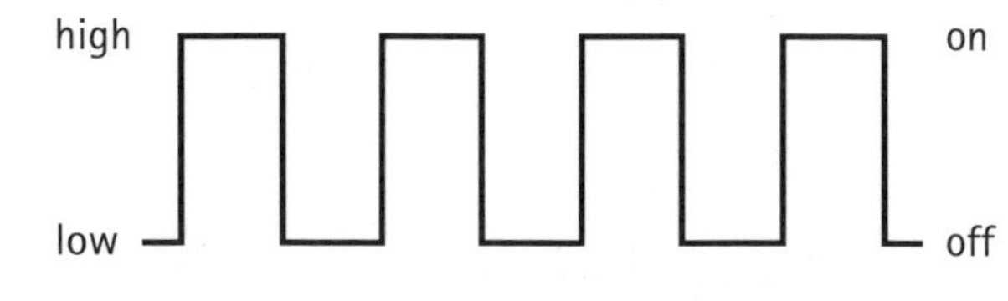

It varies in a continuous way and can have any number of in-between values.

Analogue and digital output devices

The devices in the output part of an electronic system are either analogue or digital devices. A loudspeaker is an example of an analogue output device. It converts a continuously varying electrical audio signal into sound waves. An electric motor is another example as its speed can be made to vary in a continuous way.

A buzzer is an example of a digital output device. It operates only in two states. It is either on or off. A light emitting diode is another example, as it is either switched on or it is switched off.

Light emitting diode

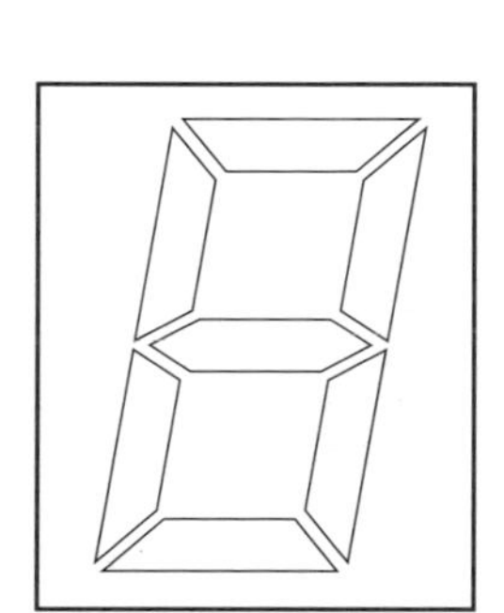

Light emitting diodes (LEDs) are used in seven segment displays. Depending on which diodes are switched on, decimal numbers from 0 to 9 can be displayed.

◎ *On the diagram, show which diodes have to be switched on to produce the number four.*

The LED has a positive end and a negative end. The LED will produce light only if it is connected correctly. The positive end of the LED must be connected to the '+' side of the supply as shown in this circuit diagram.

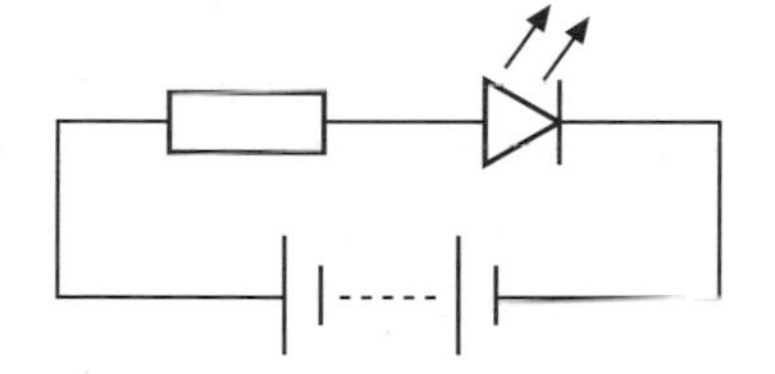

The negative end of the LED must be connected to the '–' side. An LED is normally connected in series with a resistor. The value of the resistor is chosen to ensure that the correct voltage and current are supplied to the LED. For example, in the diagram above, the LED is designed to operate normally when there is a voltage of 2 V across it and it carries a current of 2 mA.

REMEMBER Make sure you can recognise and draw the LED symbol (see page 55).

The voltage of the supply is 5 V, so the voltage drop across the series resistor is 5 - 2 = 3 V. The current in the series resistor is 2 mA. The resistance of the series resistor is obtained using the relationship:

$$R = \frac{V}{I}$$

REMEMBER At Credit Level, you need to know how to calculate the resistance of the series resistor which is used with the LED.

◎ *Complete the calculation to find the value of the resistance of the series resistor.*

$$R = \frac{3}{\square}$$

$$= \square \text{ k}\Omega$$

Electronics

Practice questions

1) What are the input and output parts of a computer system?

2) What is the process part of a hi-fi system?

Input devices

Input devices detect changes in sound levels, changes in temperature, pressure or light intensity. Some input devices operate by generating small voltages in response to changes in sound levels, temperature, etc.

A microphone generates a small voltage when it detects sound. A thermocouple produces a small voltage when it is heated. A solar cell produces a voltage when it absorbs light energy. Other input devices need a separate power supply to enable them to operate.

Can you think of a suitable input device for: a) a public address system; b) an electronic thermometer; c) a power supply in a satellite; d) controlling the switching of a security light at dusk?

When a thermistor is connected to a voltage supply, it is sensitive to temperature. The resistance of the thermistor changes as its temperature increases.

When a light dependent resistor (LDR) is connected to a voltage supply, it is sensitive to light levels. The resistance of the LDR decreases as the intensity of the light on it increases.

The change in resistance caused by changes in temperature and light level produces voltage and current variations in the input circuit. These can then be processed.

To calculate the resistance of a thermistor or an LDR you use:

$$R = \frac{V}{I}$$

An LDR in a darkened room carries a current of 0.02 amperes. The voltage across the LDR is 4 volts. What is the resistance of the LDR at this light level? Complete the calculation of the resistance.

$$R = \frac{V}{I}$$

$$= \frac{4}{}$$

$= \square$ Resistance of LDR in the darkened room $= \square$ ohms

REMEMBER At Credit Level you should know that the time taken for the capacitor to charge depends on the capacitance of the capacitor and the resistance of the resistor.

Capacitor and voltage divider

Capacitors and voltage dividers can be used to provide input voltages for electronic systems. In the capacitor circuit in Figure 1 (page 59), the voltage across the series resistor provides the input voltage to an electronic system. The capacitor produces a time delay. After the switch is closed, it may take a few seconds before the capacitor is fully charged. As the voltage across the capacitor increases, the voltage across the resistor drops gradually to zero.

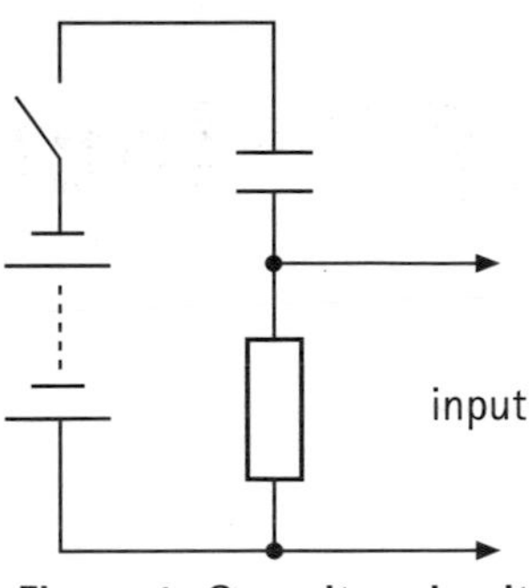

Figure 1: Capacitor circuit

A voltage divider divides up voltage. It can consist of two resistors connected in series. Alternatively, it can be a resistor connected in series with either an LDR or a thermistor (see the circuit diagram in Practice question 1, below).

Figure 2 shows a voltage divider made up of two resistors, 3 kΩ and 2 kΩ, connected across a 5 V supply.

It divides up the supply voltage like this.

The total resistance of the circuit is 5 kΩ ($R_T = R_1 + R_2$)

The current (I) in the circuit is calculated using $I = \frac{V}{R}$

$I = \frac{5}{5000} = 0.001 \text{ A} = 1 \text{ mA}$

The voltage across 3 kΩ resistor is calculated using $V = IR$

$V = IR = 0.001 \times 3000 = 3 \text{ V}$

The voltage across 2 kΩ resistor = 2 V

You can see that the voltage is divided in the same ratio as the resistance of the resistors. So if the divider had consisted of a 4 kΩ resistor and a 6 kΩ resistor, the voltages would have been:

voltage across 4 kΩ resistor $= 5 \times \frac{4}{10} = 2 \text{ V}$

voltage across 6 kΩ resistor $= 5 \times \frac{6}{10} = 3 \text{ V}$

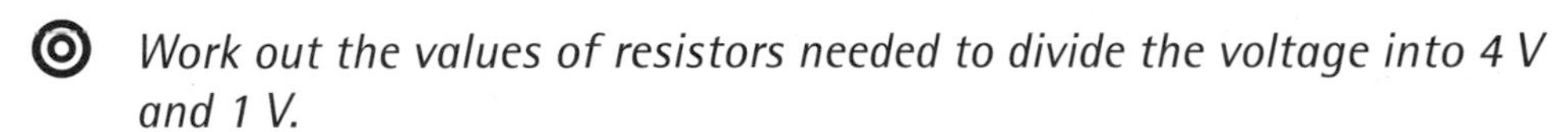

Work out the values of resistors needed to divide the voltage into 4 V and 1 V.

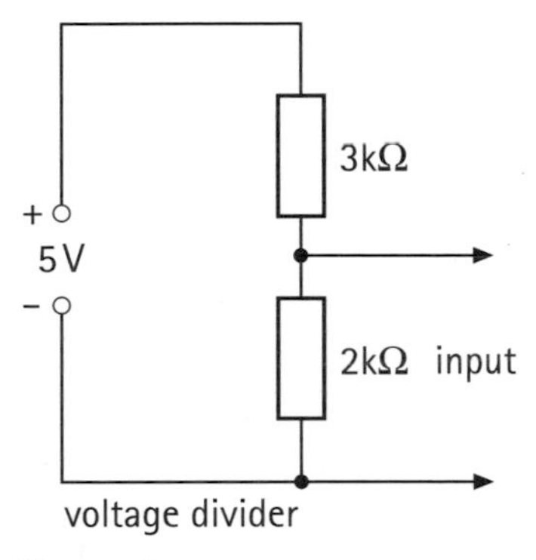

Figure 2

REMEMBER
At Credit Level, you need to be able to carry out calculations relating to voltage dividers.

Practice questions

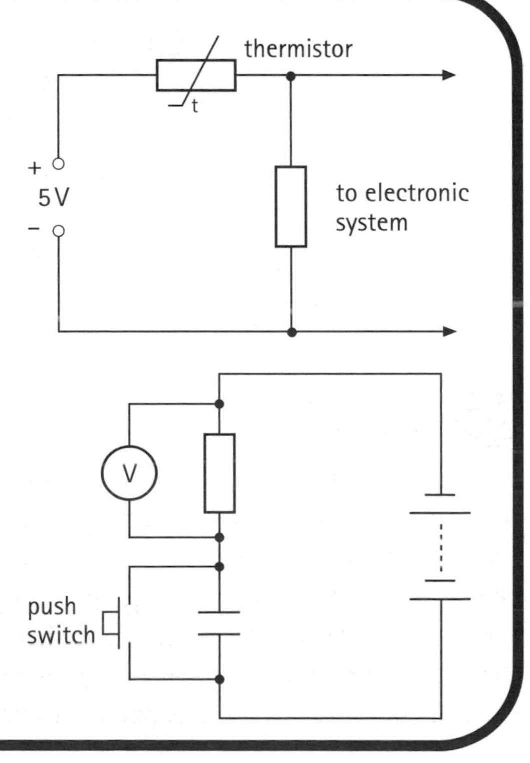

1) The input circuit shown in the diagram is designed to deliver an input of 1.25 V to an electronic system when the temperature of water in a tank reaches 100°C. The input circuit consist of a thermistor connected in series with a resistor across a 5 V supply. The resistance of the thermistor is 400 Ω when its temperature is 100°C.

   Calculate the resistance of the resistor.

2) A capacitor is used in an input circuit as shown.

   Describe what happens to the voltage across the resistor as the push switch is pressed and then released.

Processing input signals

> **REMEMBER** You need to know how to draw the symbol for a transistor.

Transistors

Transistors process the input signals. In an electronic system, the transistor acts like a switch. It is switched on by a voltage from the input section of the system.

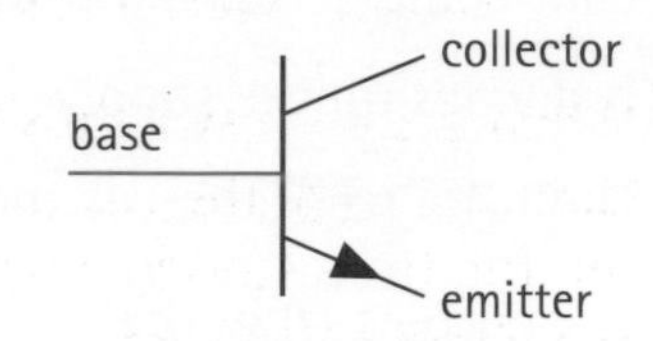

When the base of the transistor receives a positive voltage of 0.7 V or more from the input circuit, the transistor conducts. When it conducts, it acts like a closed switch.

Transistor switching circuits

The lamp in this transistor switching circuit switches on automatically when the light intensity falls to a certain level. The light level at which the transistor switches on can be set by adjusting the resistance of the variable resistor.

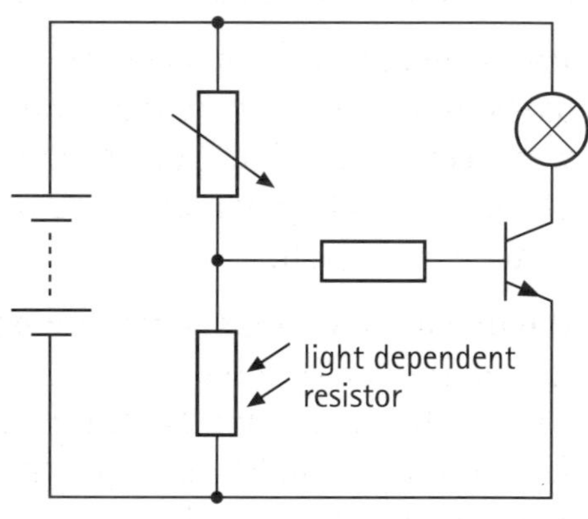

The LDR and the variable resistor form a voltage divider circuit. As the light level drops, the resistance of the LDR increases and so the voltage across the LDR increases. At a particular light level, the resistance of the LDR is such that the voltage across it is enough to switch the transistor on.

Suppose the transistor has to switch on when the resistance of the LDR is 1000 ohms. Complete this calculation to work out what resistance the variable resistor must be set at. You can ignore the small current that is supplied to the base of the transistor in carrying out the calculation.

current in LDR when transistor is switched on $= \frac{V}{R} = \frac{0.7}{1000} = 0.0007$ A

(See rule 1 for a series circuit on page 42.)

current in variable resistor = 0.0007 A

(See rule 2 for a series circuit on page 42.)

voltage across variable resistor = 5 – 0.7 = 4.3 V

resistance of variable resistor $= \frac{V}{I}$

= ——

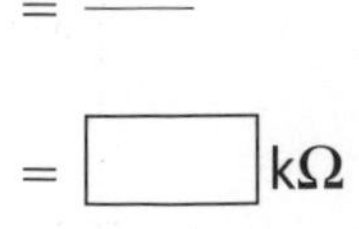
= ☐ kΩ

In what way could this circuit be used for a practical purpose?

Here is a circuit that will switch on the LED for a certain length of time only.

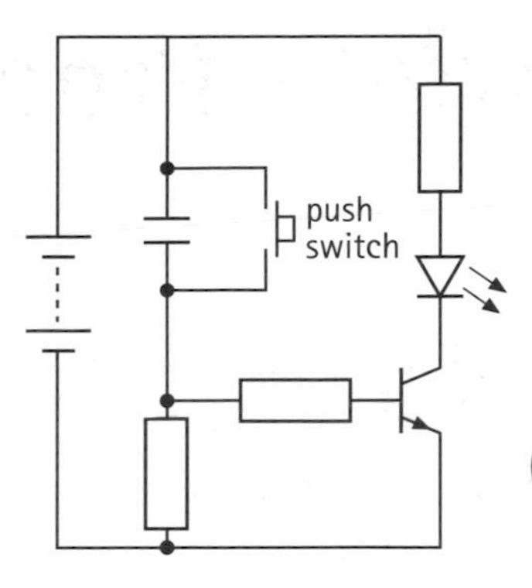

Complete the blanks in the following paragraph by using these words:

charge decreases voltage resistance conducts

When the push switch is pushed and then released, the capacitor C starts to .... up. The .... across R is large enough to switch the transistor on. The transistor conducts and so the LED is lit. As the capacitor charges up, the charging current .... The voltage across R decreases. When the voltage across R drops to less than 0.7 volts, the transistor no longer .... and the LED is switched off. The time that the LED is switched on for depends on the values of the capacitance of C and the .... of R.

Transistors and logic gates

Transistors can be connected together to make special circuits called 'logic gates'. A logic gate is a circuit that allows a voltage to appear at its output when certain combinations of high and low voltages are applied at the input of the gate. A high voltage is sometimes referred to as 'logic 1' and a low voltage as 'logic 0'.

The operation of logic gates can be summarised using tables called 'truth tables'. The tables show the output from each gate for all possible combinations of input. From the tables, you can see why the gates are called AND, OR and NOT. For example the AND gate provides an output voltage only when input A **and** input B are high.

INPUT		OUTPUT
A	B	Z
0	0	0
0	1	0
1	0	0
1	1	1

INPUT		OUTPUT
A	B	Z
0	0	0
0	1	1
1	0	1
1	1	1

INPUT	OUTPUT
A	Z
0	1
1	0

NOT

REMEMBER There are many types of logic gate – you need to know about the two-input AND gate, the two-input OR gate and the NOT gate. Make sure you can recognise and draw the symbols for each of these gates (see page 55).

Why are OR gates and NOT gates so called?

Practice question

The diagrams show how transistors are connected to make an AND gate, and OR gate and a NOT gate.

Draw the appropriate gate symbol next to each circuit.

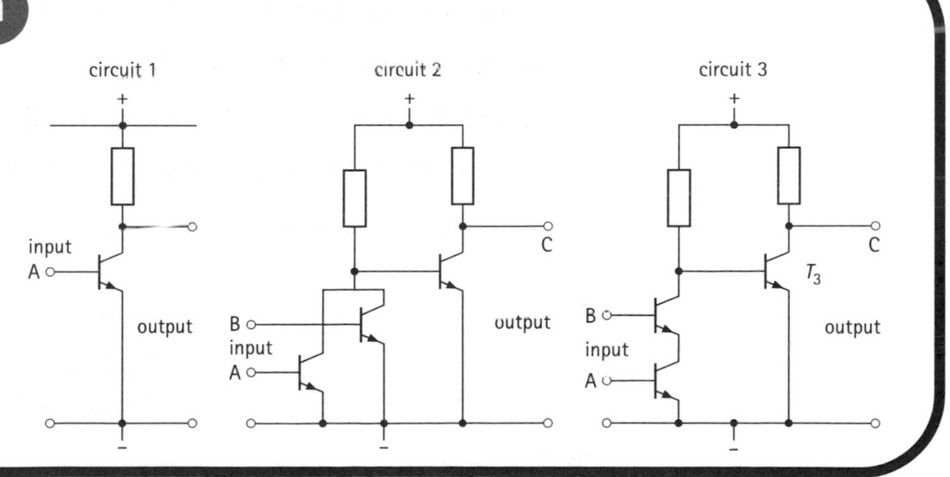

Logic gates, clocks and counters

Logic gates

Logic gates can be combined to make electronic systems that are capable of making decisions and performing certain actions.

A gate is something that lets you in or keeps you out, depending on whether it is open or shut. Input devices in electronic systems can be used to open or shut logic gates. The gate will open depending on what the inputs are and what the logic of the gate is.

REMEMBER Make sure you know how logic gates can be used in an electronic system that is being designed to perform a certain action.

Suppose you have to design a system to operate a warning buzzer if the ignition of a car is switched on and the driver has not fastened the seatbelt.

The input devices for this system are two switches: an ignition switch and a switch that is closed when the seatbelt is fastened.

The processing part of the system will be a combination of logic gates. The combination of gates will provide an output voltage when the ignition switch is on (closed) and the seatbelt switch is off (open). In the electronic system, a closed switch gives a logic 1 and an open switch gives a logic 0.

The output part of the system is a buzzer. The buzzer will sound when it is supplied with a voltage, i.e. when the output of the logic gates is high (logic 1).

Here are the block diagram and truth table for the system.

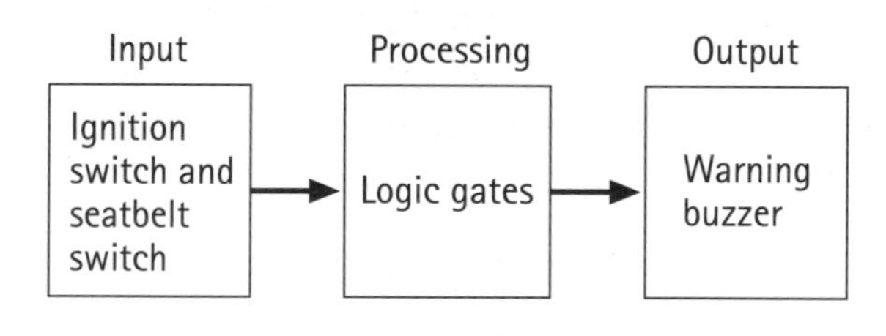

Input		
Ignition switch	**Seatbelt switch**	**Output**
0 (off)	0 (off)	0 (off)
0 (off)	1 (on)	0 (off)
1 (on)	0 (off)	1 (on)
1 (on)	1 (on)	0 (off)

You have to find the combination of gates which meets the requirements of the truth table. If you invert the logic of the seatbelt switch, so that a logic 0 becomes a logic 1 and vice versa, then the truth table looks like that of an AND gate. You can change the logic of the seatbelt switch using a NOT gate. Here is a diagram of the electronic system.

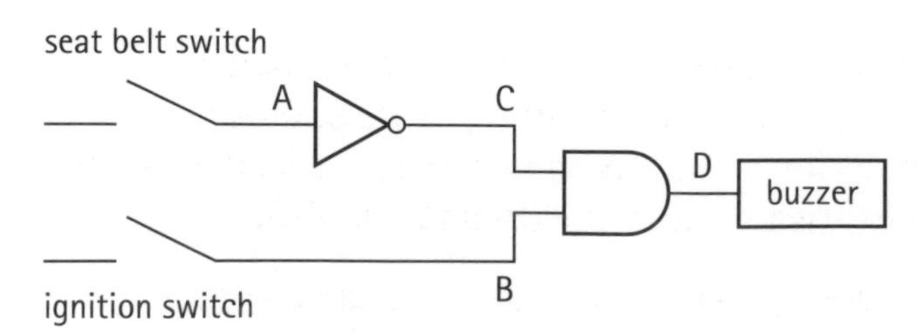

Complete the table showing the logic at points A, B, C and D in the system.

A	B	C	D
0	0	1	0
0	1		
1	0	0	
1	1		0

Clock signals

An electronic circuit can be assembled to act like a clock. Instead of ticks and tocks, the electronic circuit produces a regular sequence of voltage pulses. Clock circuits are needed in some electronic systems to make sure that certain actions are carried out at the correct time.

REMEMBER At Credit Level you have to know how the clock works.

A NOT gate, a resistor and a capacitor can make an electronic clock. The components are arranged like this. The supply voltage is omitted from the diagram.

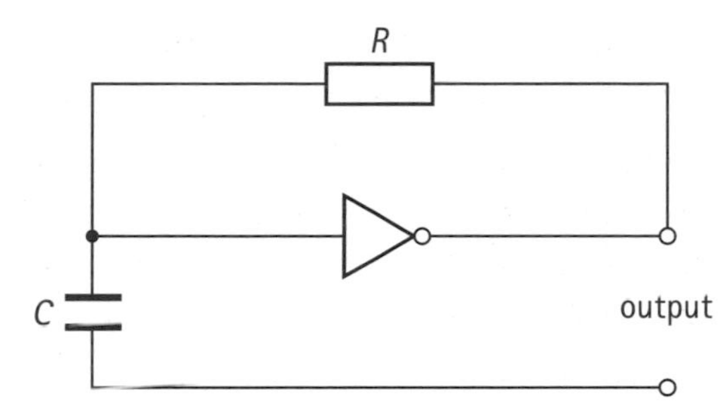

At the start of the cycle, the capacitor is about to be charged up. At this point, the output of the NOT gate is high. A current is produced in R. The current causes the capacitor to start to charge up. The voltage across it rises to a high.

The high voltage across the capacitor is applied to the input of the NOT gate. This high voltage produces a low voltage at the output of the NOT gate. The output terminals are now effectively connected together and so the capacitor is able to discharge through the resistor.

This causes the input voltage to the NOT gate to become low.

The voltage at the output of the NOT gate then becomes high. The capacitor starts to charge up again and so the oscillation continues.

The duration of a clock pulse is determined by the time taken for the capacitor to charge and the time taken for it to discharge. The frequency of the clock pulses depends on the capacitance of C and the resistance of R.

Counters

The electronic system keeps track of time by counting the clock pulses. Counting circuits use binary notation for the counting process. Binary numbers can be converted to decimal numbers using a binary decoder circuit. A seven segment display can be used to show the decimal number.

REMEMBER At Credit Level you need to be able to calculate the decimal equivalent of binary numbers in the range 0000–1001.

Which decimal numbers are represented by the range 0000–1001?

Examination questions

Try these two questions from past exam papers. Question 1 is from a **General Level** paper and question 2 is from a **Credit Level** paper. Spend about 7 minutes on question 1 and no more than 10 minutes on question 2. When you have finished, turn to page 95 for the answers and the marking guide.

1) a) A block diagram of a public address system is shown below.

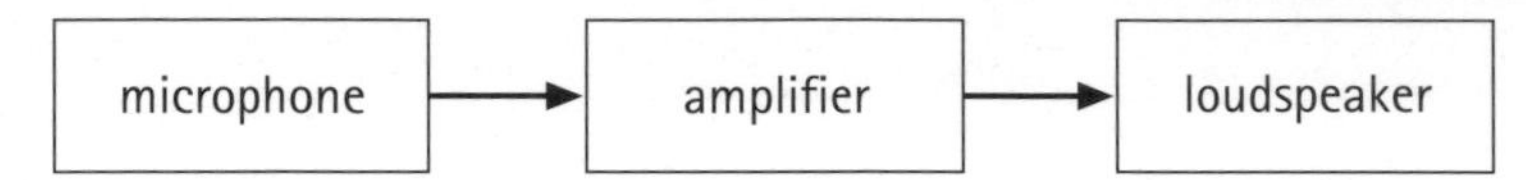

i) Name the output device. Name the input device. **(1)**

ii) State the useful energy change which takes place in the loudspeaker. **(1)**

b) The voltage gain of the amplifier is 50 000. What is meant by this statement? **(1)**

c) An input signal of 1000 hertz is applied to the amplifier. What is the frequency of the output signal? **(1)**

d) The power of the amplifier is 120 watts. How much energy is supplied to the amplifier in 5 minutes? **(2)**

C **2)** Figure 1 shows an electronic system which can be used as a timer.

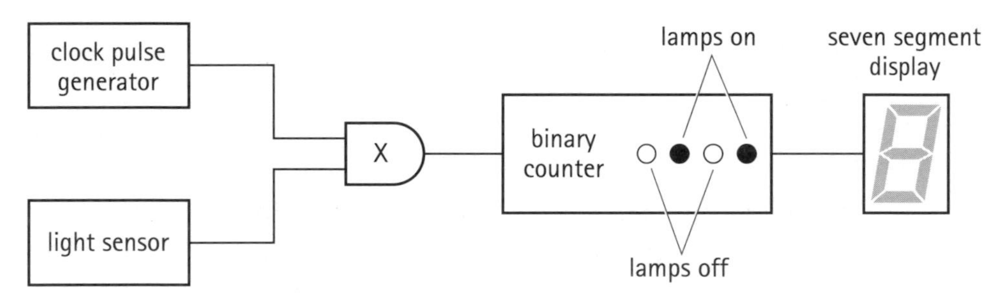

Figure 1

The timer is switched on and off using a beam of light and a light sensor.

The logic level at the output from the light sensor is shown in the table opposite.

Lighting condition at light sensor	Logic level at output from light sensor
Dark	1
Light	0

The clock pulse generator produces an output voltage which changes with time as shown at the top of the next page. The logic levels are indicated on the graph.

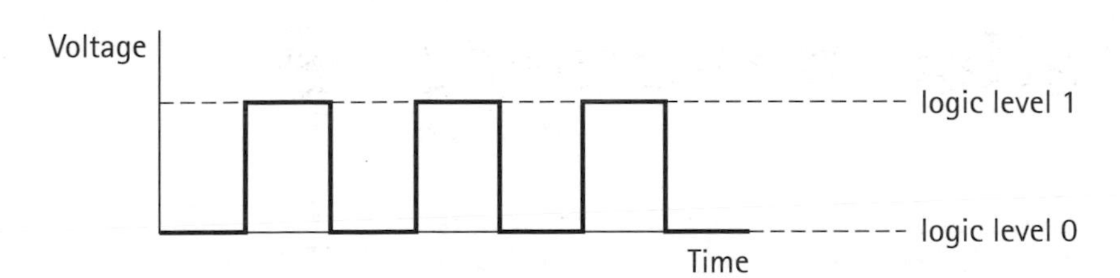

a) i) What name is given to logic gate X? (1)
   ii) Explain why no counts are recorded on the binary counter when the light beam shines on the light sensor. (2)
   iii) What number appears on the seven segment display when the binary counter displays 0101 as shown in figure 1? (1)

b) Saeed makes use of the electronic system in figure 1 to measure the speed of a toy car. He sets up the apparatus as shown in figure 2 and resets the binary counter and seven segment display to zero.

When the car has passed through the light beam between the lamp and the light sensor, the seven segment display shows the number eight. The time for each clock pulse is shown in figure 3.

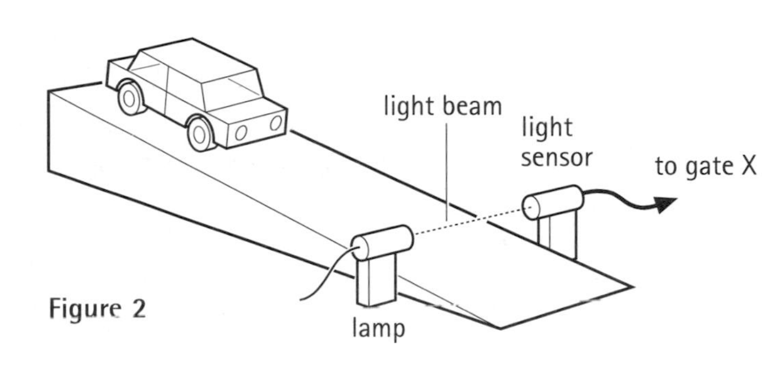

Figure 2

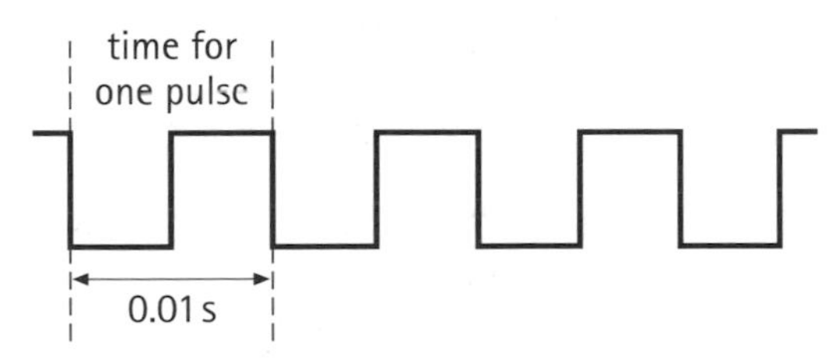

Figure 3

i) Calculate the time taken for the car to pass the light beam. (1)
ii) What other measurement is required so that Saeed can calculate the speed of the car as it passes through the light beam? (1)
iii) How would Saeed use his measurements to calculate this speed? (1)

c) The electronic circuit which produces the clock pulse is shown in figure 4.

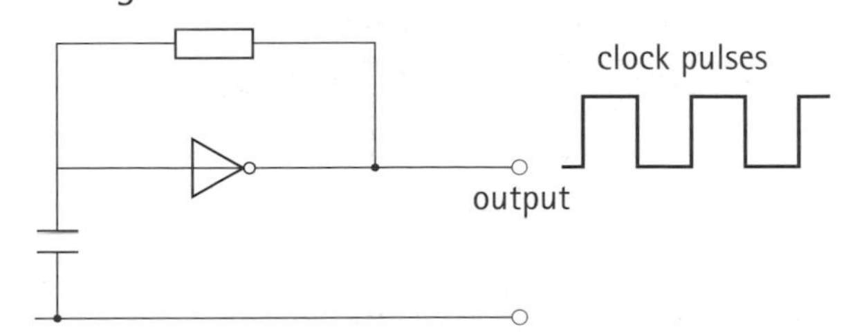

Figure 4

i) Saeed increases the value of the capacitor. What effect does this have on the frequency of the clock pulse? (1)
ii) What effect should this have on the accuracy of Saeed's time measurement? (1)

Waves

This section is about:

- using correctly the key words (in bold) below
- describing water waves, sound waves and electromagnetic waves
- describing the reflection, refraction and diffraction properties of waves and how wave properties are put to use
- calculating the speed, wavelength and frequency of waves.

All waves have common properties – they can be **reflected, refracted** and **diffracted.** Wave motion is a good way to describe how different forms of energy (e.g. sound energy, light and heat energy) are transmitted.

There are two types of waves: **longitudinal waves** and **transverse waves.** The motion of a wave can be drawn as a series of crests and troughs. The distance between two neighbouring crests or troughs is called the **wavelength** of the wave. The number of crests generated by the source of the waves every second is known as the **frequency** of the wave. The height of the crest of a wave is called the **amplitude.** The speed of the wave is the product of the frequency and the wavelength.

Sound waves and water waves depend on **particles** for their transmission. Some waves do not need particles for their transmission – an **electromagnetic wave** uses vibrating electric and magnetic fields. Electromagnetic waves are transverse waves which travel at the **speed of light.** There is a whole family of electromagnetic waves. Members of the family include **gamma radiation, X-rays, ultraviolet radiation, visible light, infrared radiation, microwaves** and **radio waves.** This family of electromagnetic waves is called the **electromagnetic spectrum.**

All these waves occur naturally and we are exposed to them daily. We can also produce some of them artificially. For example, in hospital there are machines which produce X-rays. **Lamps** can produce visible light and ultraviolet radiation. **Heaters** can produce infrared. **Electronic transmitters** can produce microwaves and radio waves.

The properties of waves can be used for different purposes. For example, the reflection of **high frequency** sound waves, called **ultrasound,** is used in hospitals to check the development of unborn babies. Members of the electromagnetic spectrum are used in **telecommunications, satellite TV, terrestrial TV, radio broadcasting** and **mobile phones.**

FactZONE

Wave quantities

speed of a wave = $\frac{\text{distance travelled by wave}}{\text{time taken}}$ $\qquad v = \frac{d}{t}$

speed of wave = frequency of wave × wavelength $\qquad v = f\lambda$

Frequency is the number of waves produced per second.

Frequency is measured in hertz (Hz), kilohertz (kHz) or megahertz (MHz).

1kHz = 1000 Hz 1MHz = 1 000 000 Hz

wavelength = distance between two neighbouring crests or troughs, measured in metres (m)

amplitude = height of the wave, from the rest position, measured in metres (m)

Sound waves

Sound travels faster in solids and liquids than in air.

Sound waves travel at a speed of around 340 m/s in air.

Sounds are carried by longitudinal waves.

Humans can hear sounds in the range of 20 to 20 000 Hz.

The louder the sound, the greater the amplitude of the sound wave. The higher the pitch of a sound, the higher its frequency.

Sound levels are measured in decibels (dB). Too high a sound level can damage hearing.

Sound beyond the range of human hearing, i.e. above 20 000 kHz, is called ultrasound.

Electromagnetic waves

Electromagnetic waves consist of vibrating electric and magnetic fields.

All electromagnetic waves travel at the same speed, i.e. the speed of light. In a vacuum, electromagnetic waves travel at 300 000 000 m/s. In glass, the speed of light waves is around 200 000 000 m/s.

The family of electromagnetic waves is called the electromagnetic spectrum.

Wave properties

Waves can be reflected. Waves obey the law of reflection.

angle of incidence = angle of reflection

Waves change speed when they move from one medium to another. The change in speed can result in a change in the direction of travel of the wave. This is called refraction.

Waves can bend round obstacles. This is called diffraction and can be observed more easily with waves of long wavelength.

Wave definitions

Transverse waves and longitudinal waves

Water waves are examples of transverse waves. In a transverse wave, the particles that make up the wave vibrate (oscillate) in a direction at right angles to the direction in which the wave is travelling.

Electromagnetic waves, such as radio and TV waves, are transverse waves. However, in these waves, it is electric and magnetic fields that vibrate, not particles.

Sound waves are examples of longitudinal waves. In a longitudinal wave, the particles of the medium that carry the wave (e.g. air) vibrate in the same direction as the wave is travelling.

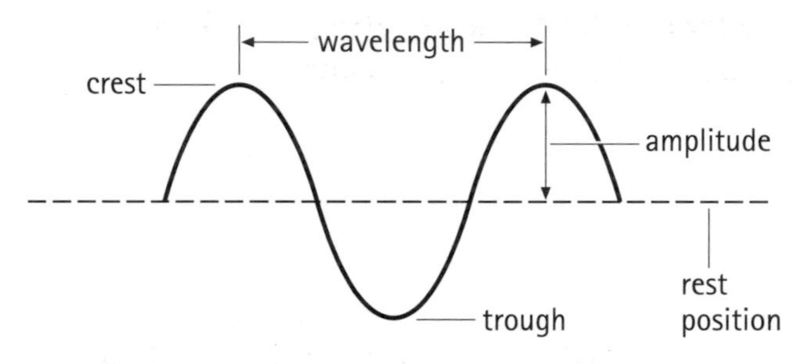

REMEMBER Whatever type of wave you are dealing with, you can draw it in the same way as you do for water waves, i.e. as a series of crests and troughs.

The **wavelength** of the wave is the distance between two neighbouring crests or troughs. Wavelength is measured in metres.

The **amplitude** of the wave is the height of a crest or the depth of a trough.

The **frequency** of a wave is the number of complete waves that are produced by the source of the waves every second. Frequency is measured in hertz (Hz). For example, a wave that has a frequency of 10 Hz means that 10 complete waves are produced every second. The frequency of waves is often described in kilohertz (kHz) (i.e. 1000 Hz) or megahertz (MHz) (i.e. 1 000 000 Hz).

Speed of a wave

The diagram you use to show a wave is like a snapshot of the wave at a particular instant in time. The diagram below shows two successive snapshots of a wave. You can see the wave is moving from left to right.

Suppose that the time between the two snapshots of the wave is 0.1 second. The crest of the wave has moved through a distance of 0.05 m in this time. You can now work out the speed of the wave.

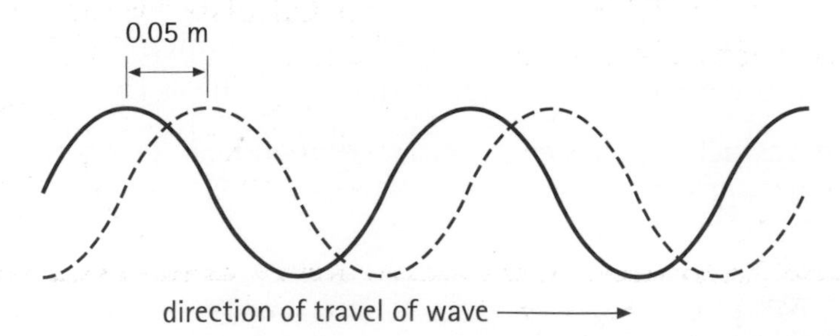

Complete this calculation.

$$\text{speed of wave} = \frac{\text{distance travelled}}{\text{time}}$$

$$= \frac{0.05}{}$$

$= \square$ m/s

Another way to calculate the speed of a wave is to use the expression $v = f\lambda$

v is the speed, measured in metres per second (m/s).

f is the frequency, measured in hertz (Hz).

λ is wavelength, measured in metres (m).

A water wave has a frequency of 4 Hz and a wavelength of 0.75 m. Complete the calculation to find the speed of the wave.

$v = f\lambda$

$v = 4 \times$ ___

$= \square$ Speed of water wave $= \square$ m/s

At Credit level, you need to know how to prove that $v = f\lambda$

Suppose a wave has frequency f. This means that f waves are produced in 1 s.

Therefore, the time taken to produce 1 complete wave must be $\frac{1}{f}$ seconds. The crest of the wave which is produced travels a distance of 1 wavelength in this time.

So speed of wave $(v) = \frac{d}{t} = \frac{\lambda}{\frac{1}{f}} = f\lambda$

Practice questions

1) Water waves produced by a wave machine in a swimming pool travel 25 m in 5 s. The wave machine produces 4 waves a second.
   a) What is the frequency of the waves?
   b) Calculate the speed of the waves.
   c) Calculate the wavelength of the waves.

2) The diagram shows a water wave. The frequency of the wave is 3 Hz.
   a) What is the amplitude of the wave?
   b) What is the wavelength of the wave?
   c) Calculate the speed of the wave.

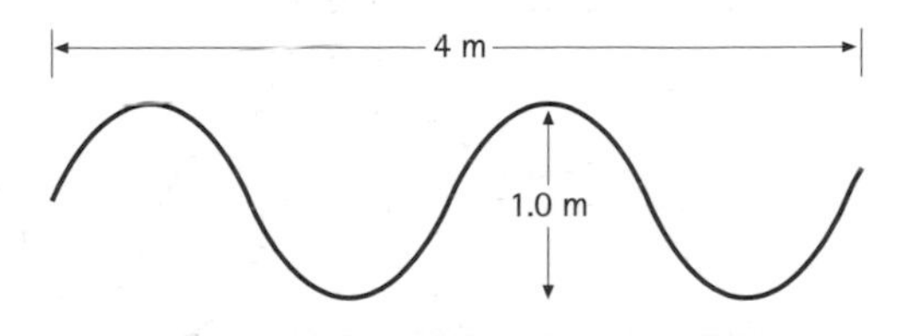

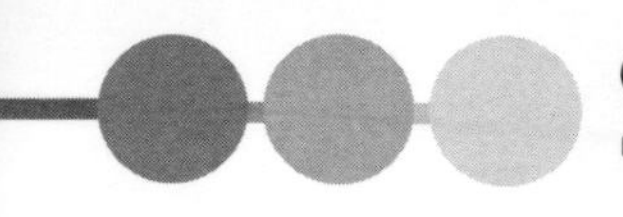

Sound waves

Sound is produced by vibration. When a guitar string is plucked, the string vibrates. The vibration of the string transfers energy to the air particles. The air particles are made to vibrate and a longitudinal wave is produced in the air. When this sound wave arrives at your ear, it makes your eardrum vibrate. The vibration of your eardrum causes electrical signals to be sent to your brain. Your brain interprets these signals as sound.

Sound needs a medium (i.e. a solid, liquid or gas) to enable it to be transmitted. Sound cannot be transmitted in a vacuum.

Speed of sound

REMEMBER High frequency high pitch sound looks like this:

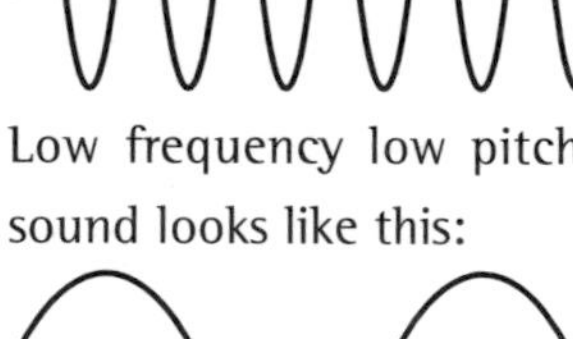

Low frequency low pitch sound looks like this:

You can measure the speed of a sound wave in air by timing how long it takes to travel a measured distance between two microphones. The microphones are connected to an electronic switch and timer.

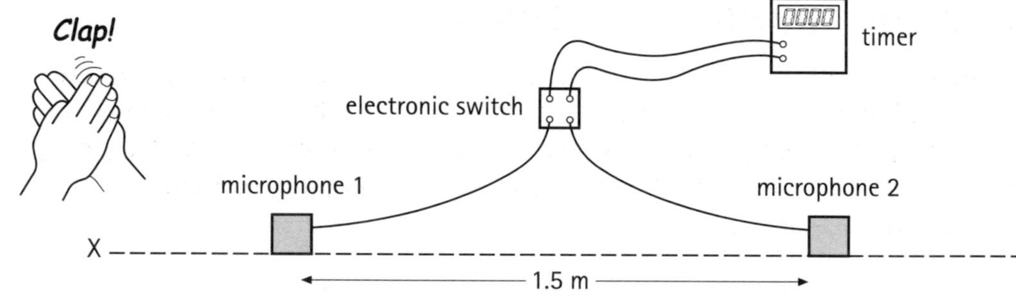

You clap your hands once so that a sound is produced. When the sound from the clap reaches the first microphone, the signal produced by the microphone triggers the timer. The timer starts timing.

When the sound reaches the second microphone, the signal produced makes the timer stop. The reading on the timer is the time taken for the sound wave to travel the measured distance.

REMEMBER Make sure you can describe what happens to the signal pattern on an oscilloscope screen when the loudness and pitch of a sound is changed.

Here is a set of readings obtained from the experiment. Use them to calculate a value for the speed of sound in air.

distance between microphones = 1.5 m

reading on timer = 4.4 ms

$$\text{speed of sound in air} = \frac{\text{distance}}{\text{time}}$$

$$= \frac{1.5}{4.4 \times 10^{-3}}$$

$$= \square \text{ m/s}$$

Sound levels

Loud sounds can be annoying. There are legal limits to the amount of noise we can be expected to tolerate. Any more than that is called 'noise pollution'. Exposure to high levels of sound over a long time can be harmful to hearing. Sound levels can be measured using a sound level meter. Sound levels are measured in decibels (dB).

Here are some typical sound levels:

silence	0 dB
a whisper	20 dB
normal conversation	60 dB
a few metres from a pneumatic drill	100 dB

High sound levels can actually be painful, a sound level of around 130 dB would be almost unbearable.

The sound level depends on the amplitude of the sound wave. The greater the amplitude of the wave, the greater the sound level.

Uses of sound in medicine

A doctor can check if a patient is healthy by using a stethoscope to listen to the sound produced by the patient's heart and lungs. The detector end of the stethoscope has two parts to it. One part is used for detecting high frequency sounds from the lungs and the other is used for detecting low frequency sounds from the heart.

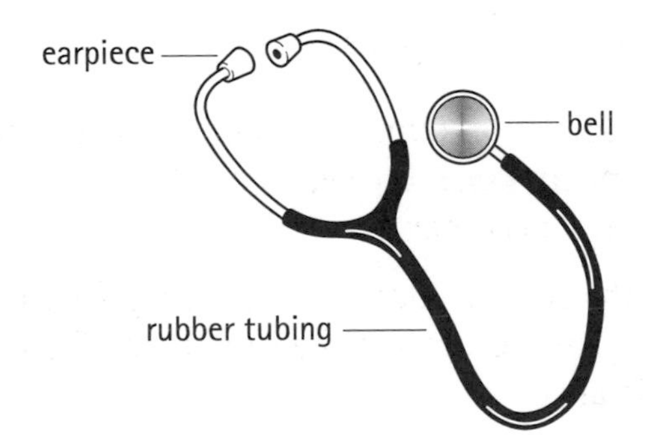

Ultrasound is also used in medicine. Doctors use ultrasound to scan a baby in its mother's womb to check its development. Ultrasound from a transmitter produces echoes in the same way as audible sound does. The echoes reflected from the baby's body are detected and processed by a computer to form an image on a screen.

Practice questions

1) A whistle produces a sound of frequency 1360 Hz. The speed of sound is 340 m/s. What is the wavelength of the sound?

2) Ultrasound of frequency 50 kHz travels 3 km in water in a time of 2 s. Calculate the speed of the ultrasound in water.

The electromagnetic spectrum

Electromagnetic waves

As their name suggests, these waves have something to do with electricity and magnetism. Electromagnetic waves are different from sound waves and water waves in that they do not have any vibrating particles. Instead, they are made up of vibrating electric and magnetic fields. Another difference is that electromagnetic waves do not need a medium (i.e. solid, liquid or gas) to enable them to be transmitted. They can travel though a vacuum. All electromagnetic waves travel through a vacuum at the same speed, the speed of light (300 000 000 m/s).

Electromagnetic waves transmit visible light. This is the radiation you can detect with your eyes. Visible light is one member of a whole family of electromagnetic waves. The family of electromagnetic waves is called the electromagnetic spectrum.

TV and radio waves

REMEMBER Make sure you know how radio and TV waves are used for communication. Make sure you can draw diagrams to show how they are produced and how they are detected.

Beyond the waves that form the visible part of the electromagnetic spectrum are longer wavelength electromagnetic waves. These waves can be used for TV and radio transmissions. Satellite TV uses electromagnetic waves called microwaves with wavelengths of a few centimetres. Terrestrial television uses longer wavelength electromagnetic waves. Local radio, police and ambulance radio communications use electromagnetic waves with wavelengths of a few metres. Worldwide radio broadcasting can use wavelengths as long as 1500 m.

Radio transmitter

One way of transmitting radio signals using electromagnetic waves is by amplitude modulation (AM). In a radio transmitter that uses AM, a high frequency electrical signal is combined with the lower frequency audio signals from a microphone. The combined electrical signal is fed to an aerial. The aerial transmits the signal as a type of electromagnetic wave – a radio wave. The radio wave carries audio information and is called a 'carrier wave'.

Radio receiver

The diagram shows the main parts of a radio receiver.

1 The aerial of the radio receiver detects radio waves and turns them into high frequency electrical signals.

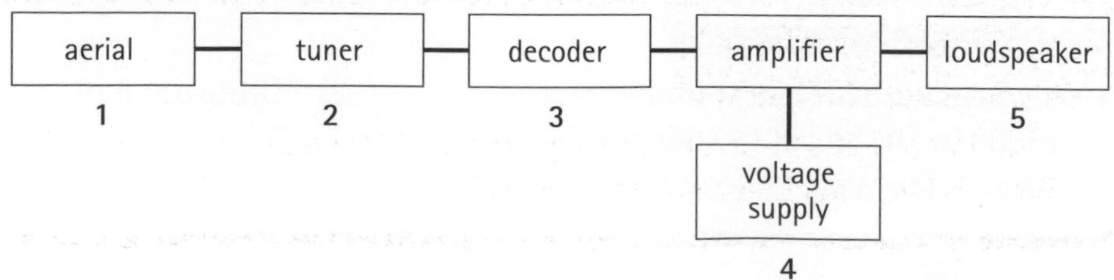

2 The tuner selects the electrical signals that come from the radio station whose broadcast the listener wants to hear.

3 The decoder filters out the audio part of the high frequency electrical signal and passes this to the amplifier.

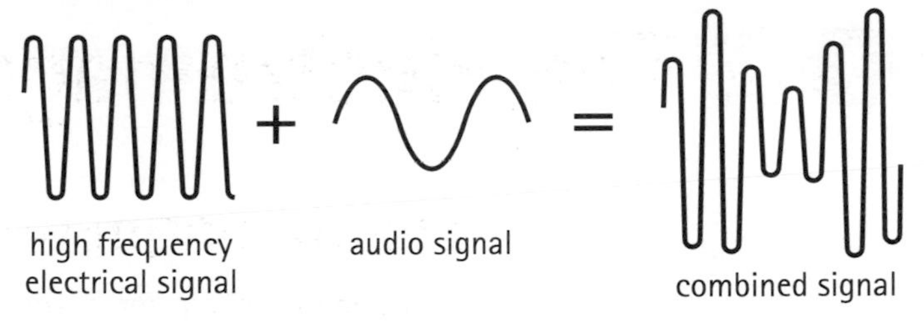

4 The amplifier boosts the audio signal and passes the boosted signal to the loudspeaker. The voltage supply provides the energy for the amplified audio signals.

5 The loudspeaker converts the electrical audio signals into sound waves.

TV transmission and reception

TV transmission is similar to radio transmission. However, in TV transmissions, video **and** audio information are carried by the electromagnetic waves. The TV waves are detected by the aerial of the TV receiver in a similar way to radio waves.

The main parts of a TV receiver are shown in the diagram. Label the parts of the diagram which deal with the audio signal.

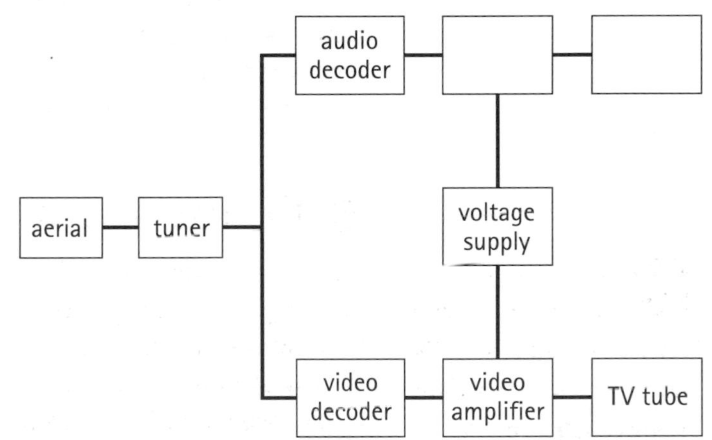

The picture on the TV tube is produced by an electron beam scanning a phosphor-coated screen. The scanning beam produces lines on the screen. Each line is a slice of the complete TV picture.

The electrons in the beam strike the screen and cause the phosphors to emit light. The video signal controls the numbers of electrons that strike the screen. This causes variation in the brightness of the light coming from the screen. When the beam makes a complete scan of the screen, all of the lines build up a picture on the screen.

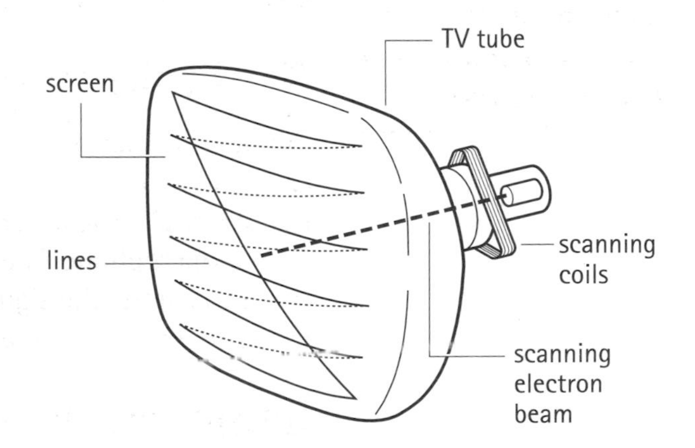

Practice question

A radio station uses a carrier frequency of 200 kHz.
What is the wavelength of the carrier?

REMEMBER Make sure that you know about the use of visible radiation, X-rays, ultraviolet radiation, infrared radiation and gamma radiation in medicine.

Electromagnetic waves in medicine

Visible light produced by a laser can be used in surgery. The light from an argon laser can be used to remove marks from the skin. Lasers are used in eye surgery to correct vision defects.

X-rays are electromagnetic waves that can penetrate body tissue. When X-rays pass through the body, the bone absorbs more of the X-rays than the flesh. If a photographic plate is placed behind the body, it can be used to detect the X-rays passing through. The X-rays produce a kind of 'shadow' photograph. The solid bone which prevents most of the X-rays reaching the film, shows up as a white 'shadow' on the film. Dark areas are produced where the X-rays reach the film. The fracture in a bone shows up as a dark region. X-rays are useful for diagnosing fractures and monitoring them as they heal.

Too much exposure to **ultraviolet radiation** may produce skin cancer. However, a controlled amount of ultraviolet radiation on the skin can be beneficial. Skin disorders such as acne can be treated by controlled exposure to ultraviolet radiation.

When **infrared radiation** is absorbed by body tissue, the tissue heats up. The heating effect of infrared radiation is used to treat damaged muscles. The heat increases the flow of blood to the damaged tissue and speeds up the healing process.

Infrared radiation can also be used to diagnose tumours. A tumour emits more infrared than healthy tissue. An infrared image of the body, called a thermogram, is produced. The thermogram is then used to find the location of the tumour.

Electromagnetic waves and astronomy

Visible radiation reaching us from stars can be passed through a spectroscope. This produces a line spectrum of the source of the visible radiation. The lines of colour in the spectrum give information about the characteristics of the atoms of the elements that are present in the source. The line spectrum is a clue to the elements present in the source of the radiation. As well as visible radiation, radio waves, microwaves and X-rays give astronomers information on what stars and galaxies are made of and how the universe has evolved.

Radio telescopes with large dish aerials have been built to detect radiation from space in the radio frequency part of the electromagnetic spectrum. The telescope at Jodrell Bank in Cheshire is a famous example of a radio telescope.

Other telescopes have been built to make their observations from orbiting satellites. Telescopes that are in orbit are not affected by the Earth's atmosphere, which acts as an absorber for some of the electromagnetic radiation from space. Therefore, telescopes in orbit are able to detect sources of electromagnetic radiation that would not be detected by telescopes on Earth.

Members of the electromagnetic spectrum

Electromagnetic wave	Examples of sources of waves	Approximate wavelength	Examples of detectors	Some uses
Radio (includes waves used for television transmissions)	Transmitters; outer space	1 kilometre to 1 metre (1 km–1 m)	Aerial and electronic circuit	Communications: radio (LW, MW, SW, VHF) and television (UHF); astronomy
Microwaves	Transmitters; outer space	A few centimetres (10^{-1} m)	Aerial and electronic circuit	Communications satellites; telephony; ovens
Infrared	Electronic devices; warm objects; Sun	A fraction of a millimetre (10^{-4} m)	Special film; thermocouple; thermistor	Security systems; remote control for television set
Visible light (red, orange, yellow, green, blue, indigo and violet)	Stars and Sun; lamps; light emitting diode	A fraction of a millionth of a metre (0.0007–0.0004 mm)	Eye; photographic film; light dependent resistor	Sight; photography
Ultraviolet	Fluorescent tubes; very hot objects; Sun	A hundredth of a millionth of a metre (10^{-8} m)	Film; fluorescent material	Sun-tan lamps; production of vitamin D
X-rays	X-ray machines; outer space	A thousandth of a millionth of a metre (10^{-9} m)	Film	Checking for defects in metals; airport baggage checks
Gamma rays	Radioactive materials; Sun	A millionth of a millionth of a metre (10^{-12} m)	Film; Geiger-Müller tube	Medical tracers; killing bacteria

Practice questions

1) What is meant by LW, MW and SW? What do VHF and UHF mean?

2) In a thunderstorm, the thunder and the lightning are produced at the same time. Why is it you see the lightning flash some time before you hear the sound of the thunder?

3) Figure 1 shows the line spectrum of a star. Figure 2 shows the spectral lines of three elements X, Y and Z.

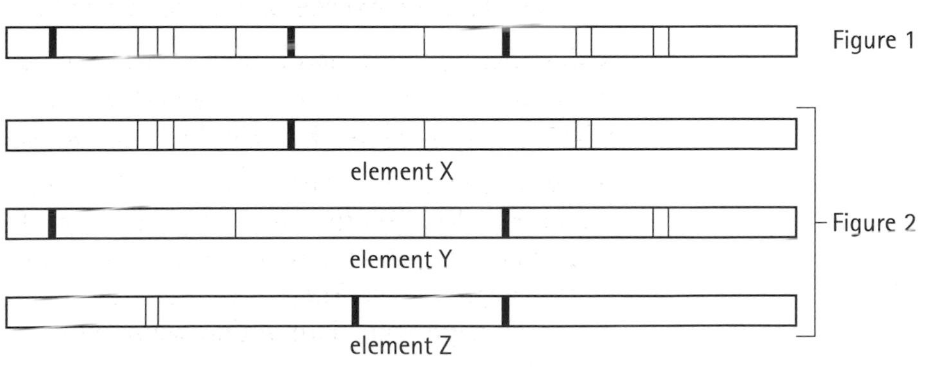

Which of the elements X, Y and Z are present in the star?

Properties of waves

REMEMBER All waves, whether water waves, sound waves or electromagnetic waves, have similar properties. Waves can be reflected, refracted and diffracted.

Reflection

Water waves can be reflected from a barrier. Light waves can be reflected from a mirror. Electromagnetic waves can be reflected using the curved dish in a dish aerial.

When light rays reflect from a mirror, they obey the laws of reflection. One of the laws of reflection says that the angle at which a light ray strikes a mirror is equal to the angle at which it is reflected. The law is often stated simply as:

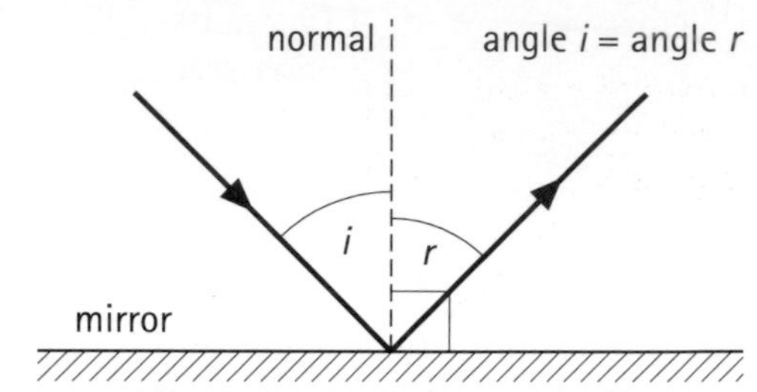

angle of incidence = angle of reflection

The curved reflector of a dish aerial collects the TV waves that bypass the detector. The dish reflects the waves and they are brought to a focus at the detector. The reflection of the waves back to the detector increases the strength of the signal in the detector.

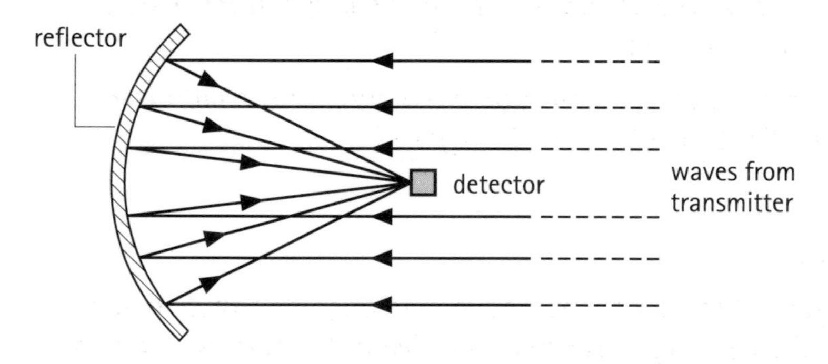

Light can be transmitted along an optical fibre using a process called total internal reflection. An optical fibre is a very thin, flexible rod of pure glass through which light can be transmitted at a speed of about 200 000 000 m/s – about two-thirds the speed of light in a vacuum (or air). Repeated internal reflections enable the light to be transmitted along the length of the fibre.

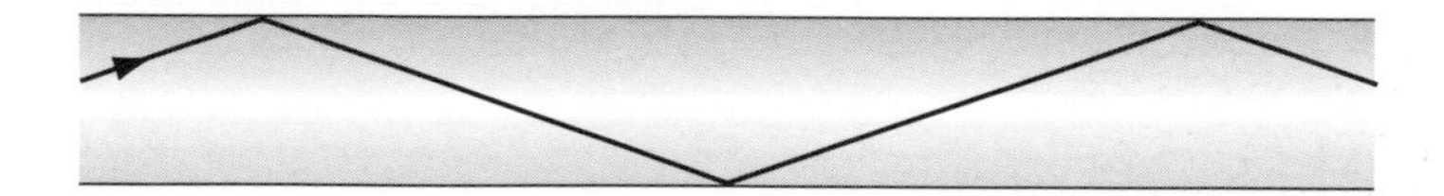

Optical fibres can be used in telecommunication systems. For example, the electrical audio signals from a telephone can be changed into light signals which are then transmitted along a length of optical fibre. The light signals are transmitted along the fibre at very high speed. The light in the optical fibre is detected at the receiving end of the system by a photodiode. The photodiode changes the light signals back into electrical, audio signals. The audio signals are then changed into sound waves.

Refraction

When waves travel from one medium to another, a change in the speed of the waves can occur. The change in speed can give rise to a change in the direction in which the wave is travelling. For example, when light travels from air to glass, the light waves travel more slowly in the glass. If the light strikes the glass-air boundary at an angle, the waves change direction as they move through the glass. The waves also change direction in travelling from glass to air.

refraction of light travelling from air to glass

i

angle of incidence

air

glass

r

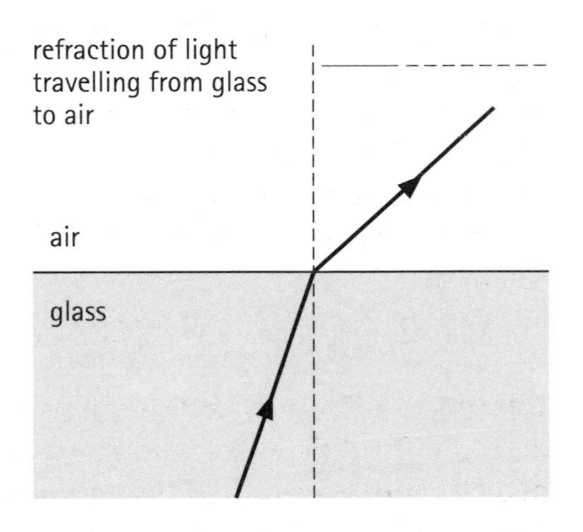

Complete the labels on the diagram to show the angle of incidence, angle of refraction and normal.

Under what conditions would a change in direction of the light ray not take place, when light travels from air to glass?

> **REMEMBER** At Credit Level, you should be able to use correctly the terms 'angle of incidence', 'angle of refraction' and 'normal'.

> **REMEMBER** Make sure you can draw a diagram to show refraction of a wave.

Diffraction

Water waves can actually bend round obstacles. The longer the wavelength, the more noticeable the bending. This effect is called diffraction. Sound waves and electromagnetic waves also can be diffracted. Sound waves can diffract round corners. Radio waves can diffract round obstacles. Diffraction of radio waves enables them to be detected even in areas where the receiving aerial is surrounded by hills. Long wavelength and medium wavelength radio diffract more than the shorter wavelength VHF waves, so radio reception with these waves is possible in most locations.

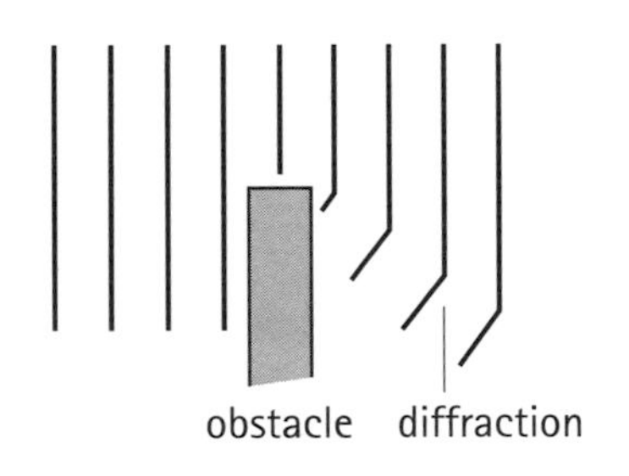

The aerial of a portable TV set can still receive TV signals even when the aerial of the TV set is not in direct line of sight with the transmitter. Mark the parts in this diagram where the signal strength is zero, weak, strong.

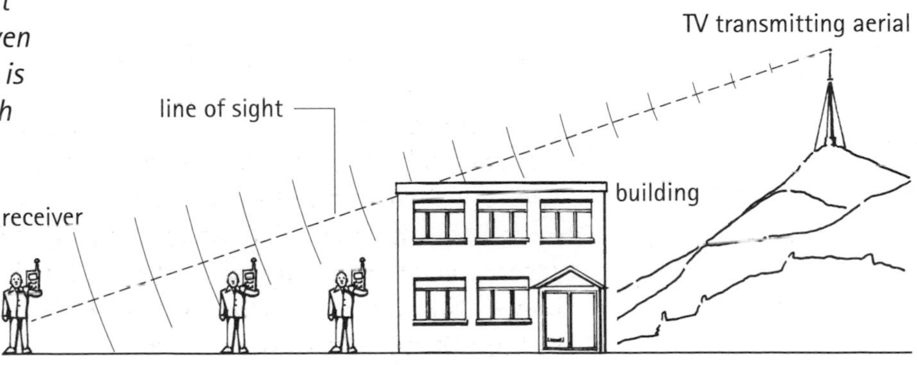

Refraction and lenses

Concave and convex lenses

REMEMBER At Credit Level, you should know that the power of a lens $= \frac{1}{\text{focal length of the lens}}$ where the focal length is measured in metres.

Convex and concave lenses refract light rays that are parallel to the axes of the lenses as shown.

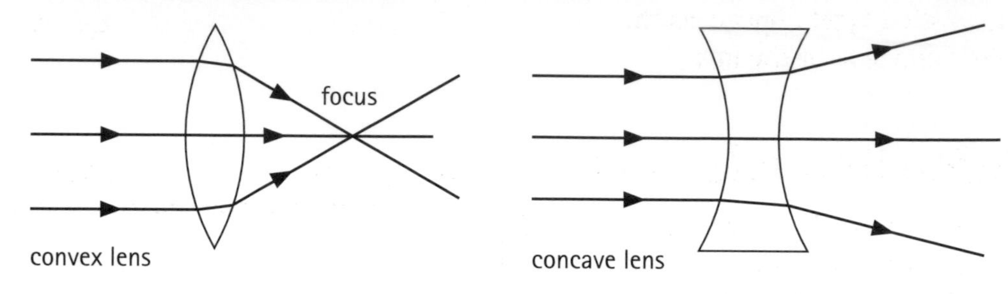

In the convex lens, the refraction causes the rays to be brought to a focus. In the concave lens, the rays diverge outwards from the focal point of the lens.

REMEMBER Lens power is measured in dioptres (D), focal length is measured in metres (m).

Where do the diverging rays in the concave lens appear to originate from? Using a ruler, project the rays backwards. Find the point on the diagram from which the rays appear to start.

A bigger curvature lens, i.e. one with a small focal length, refracts the light more than a lens with smaller curvature. The lens with a big curvature is more powerful.

Lenses and images

A lens can form an image of an object. When rays of light from a point on an object pass through a convex lens, they are refracted according to two rules.

Rule 1 A ray parallel to the axis of the lens is refracted through the focal point of the lens.

Rule 2 A ray that is directed at the centre of the lens passes straight through the lens.

These two rules can be used to find the position of the image that is formed by the lens.

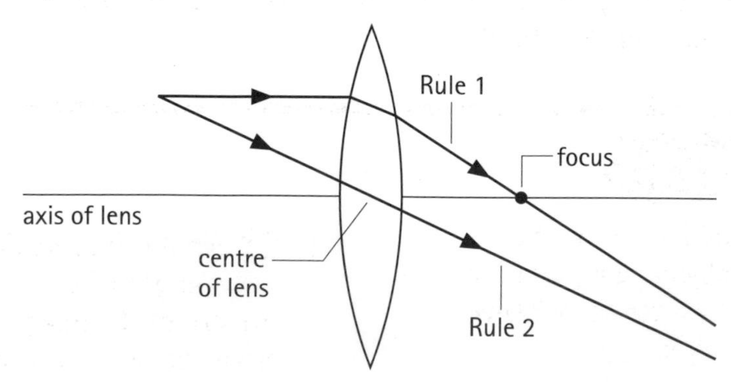

The lens of the eye is a convex lens. You can use a ray diagram to show how the eye lens forms an image of the object on the retina.

The image formed is upside down compared to the object. It is also laterally inverted, i.e. the left-hand side of the object appears as the right-hand side of the image.

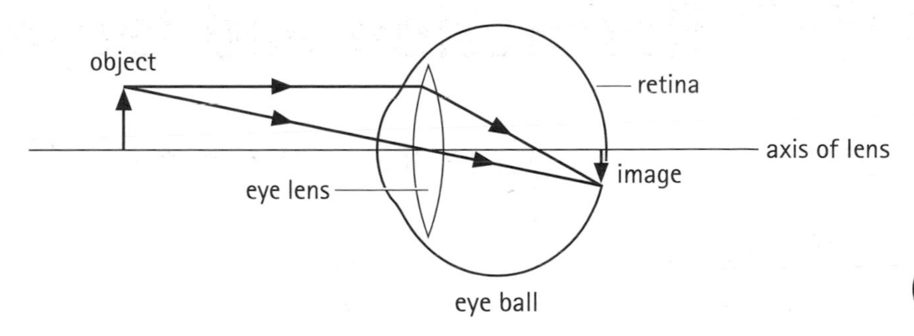

Refracting telescope

The refracting telescope uses the refracting property of lenses. The telescope consists of a sliding tube.

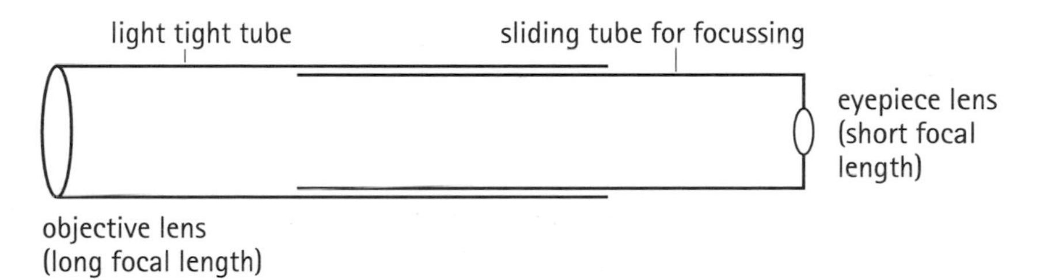

On one end of the tube is a long focal-length convex lens. This is called the objective lens. At the other end is the shorter focal-length eyepiece lens. The sliding tube enables you to bring into focus the image of the object that is being viewed.

The objective lens forms an image of the object that is being viewed. This image acts as the object for the eyepiece lens. The eyepiece lens produces a magnified image.

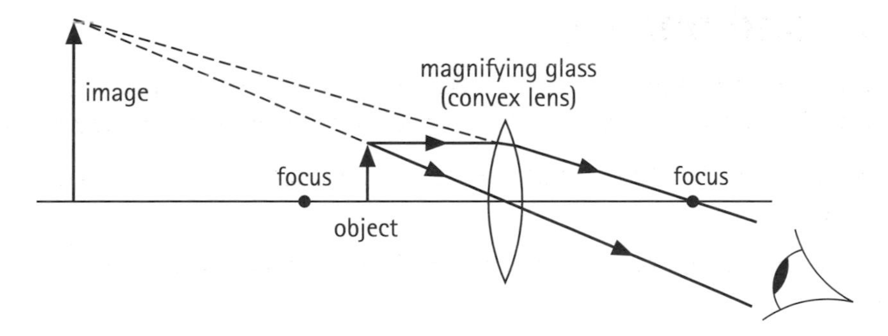

REMEMBER At Credit Level, you need to know how to draw a ray diagram to show how a convex lens acts like a magnifying glass and produces a magnified image of an object.

In order for a convex lens, such as the eyepiece of a telescope, to act like a magnifying glass, the object being magnified must be placed a distance from the lens that is less than the focal length of the lens.

Practice questions

1) What effect would increasing the diameter of the objective lens have on the image produced by a telescope? Give a reason for your answer.

2) C What effect would increasing the focal length of the eyepiece lens have on the size of the image (magnification) produced by a telescope?

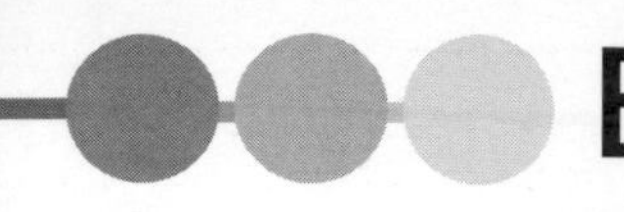

Examination questions

Try these two questions from past exam papers. Question 1 is from a **General Level** paper and question 2 is from a **Credit Level** paper. Spend about 5 minutes on question 1 and no more than 9 minutes on question 2.

When you have finished, turn to page 96 for the answers and the marking guide.

1) The diagram shows children using a large float in the swimming pool of a sports complex. A wave machine in a swimming pool generates 24 waves per minute on the surface of the pool.

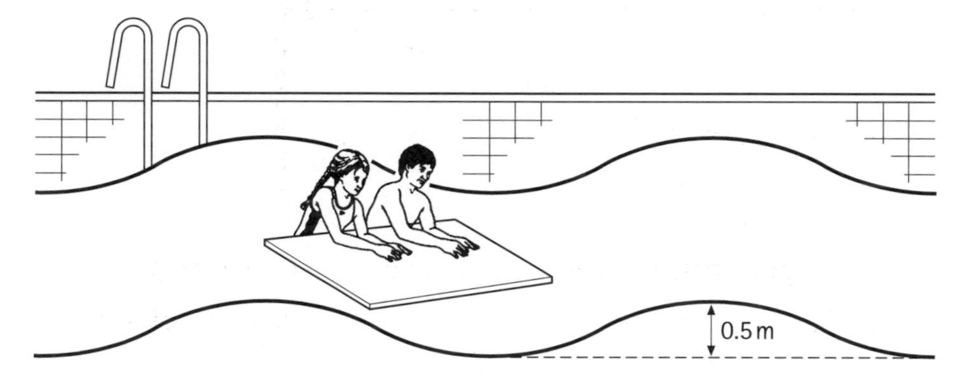

a) Show that the frequency of the wave machine is 0.4 hertz **(2)**

b) The wavelength of the waves in the pool is 4.0 metres. Calculate the speed of the waves in the pool. **(2)**

c) The large float moves up and down on the waves. The vertical distance through which the float rises is 0.5 metres. What is the amplitude of the waves? **(1)**

C **2)** In the eye, refraction of light takes place at the cornea and the eye lens.

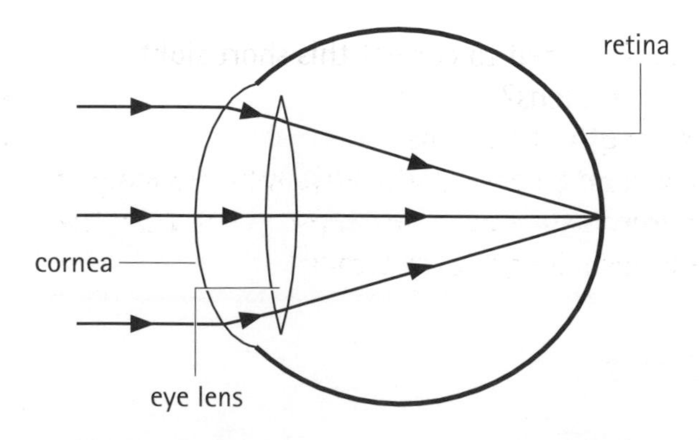

a) What is meant by refraction of light? **(1)**

b) More of the refraction takes place at the cornea as shown on page 80 (opposite). To show how the eye forms an image, a student uses two identical lenses and a screen to make a model eye. Three parallel rays of light are directed towards the lenses and are focused on the screen as shown.

State one change that could be made to the lens system to represent the eye more correctly. Explain your answer. (2)

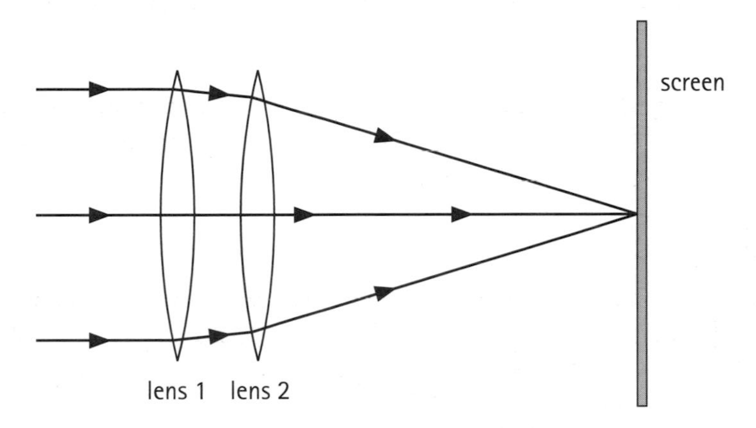

c) The diagram below represents the eye of a short-sighted person.

i) On the diagram, circle the letter where all three rays shown would meet. (1)

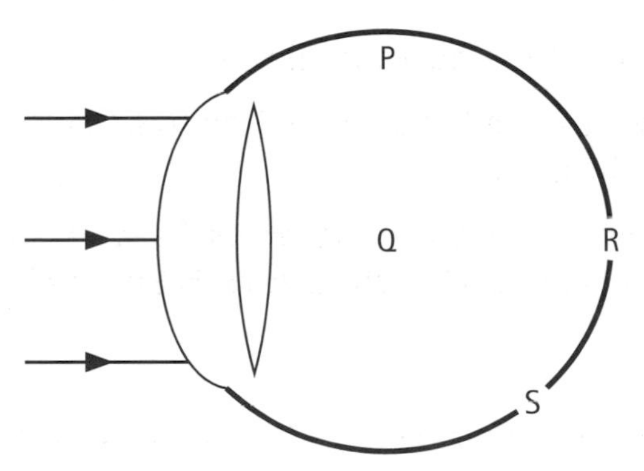

ii) A lens of power −4.0 D is used to correct this short sight. What is the shape of this lens? (1)

iii) Calculate the focal length of the lens. (2)

iv) Laser surgery can be used to correct short sight by reducing the curvature of the cornea in the eye. What effect would the laser surgery have on the focal length of a cornea? (1)

Radioactivity

This section is about:

- using correctly the key words (in bold) below
- describing the properties of alpha, beta and gamma radiation and how these radiations are detected
- describing how radiation can be handled safely and used in medicine
- calculating the half-life of a radioactive source.

Certain elements emit **radiation**. These elements are known as **radioactive elements**. Three types of radiation are emitted by radioactive elements: **alpha, beta** and **gamma** radiation. Some radioactive elements occur naturally. Others are man-made. Every day our bodies are exposed to **background radiation**. This can come from naturally occurring **radioactive rocks**. Exposure to low levels of radiation from radioactive sources is part of normal everyday living.

When working with radioactive materials, it is important to take **safety precautions** and to limit the **dose** of radiation received from the radioactive source. Radiation doses are measured in **sieverts**. Radiation produces **ionisation**. This can disturb the way the cells in our bodies operate and can cause the cells to become cancerous.

When radiation is used in a **careful and controlled** way, it can be beneficial. Radiation is used in medicine. For example, gamma radiation from a radioactive source can be directed into the body so that the radiation targets and destroys cancerous cells.

Radioactive sources, such as iodine –123 and technetium –99, can also be used by doctors to find out if the body is working properly. Doctors can inject a small amount of radioactive material into a patient's body. The radioactive material is carried by the blood to the part of the body the doctor wants to examine. The radiation coming from the affected part is measured using special detectors and electronic circuitry. The measurements are processed by a computer so that a clear image of the affected part of the body is displayed on the computer screen. The image helps doctors to **diagnose** what is wrong.

The activity of a radioactive source is measured in **becquerels**. The activity of a radioactive source reduces as time passes, so the number of radiations emitted from the source every second decreases. The **half-life** of the source is the time taken for the activity of a source to reduce by half. Knowing the half-life of a radioactive source allows technicians to calculate what the activity of source will be at a particular time and when it can be used safely with a patient.

Properties of radiation

Type of radiation	Nature of radiation	Speed of radiation	Absorption of radiation
alpha (α)	Helium nuclei	5% of speed of light	Stopped by paper
beta (β)	Electrons	90% of speed of light	Stopped by a few millimetres of aluminium
gamma (γ)	Electromagnetic waves	Speed of light	Stopped by many centimetres of lead or a few metres of concrete

Background radiation

This is the radiation that is present all around us, which we are exposed to as part of our everyday living. Most of this radiation comes from naturally-occurring radioactive sources.

Detecting radiation

Radiation can be detected by a Geiger-Müller (GM) tube.

The radiation entering the GM tube can be counted. A count rate (counts per second or counts per minute) is measured. When measuring the count rate from a radioactive source, you have to correct for background radiation.

count rate from radioactive source = measured count rate – background count rate

Ionisation

Here, electrons are stripped from the atoms in the absorbing material. Alpha, beta and gamma radiation produce different amounts of ionisation. For example, in passing through a certain thickness of air, alpha radiation will produce most ionisation, beta next and gamma least.

Safety and dose

To limit the radiation dose from a radioactive source, you can handle it from a distance, shield yourself from it and limit the time you are exposed to the radiation.

The quantity 'dose equivalent' tells you how much of a dose has been received. Dose equivalent is measure in sieverts (Sv).

Activity and half-life

The activity of a source is the number of radiations being emitted by the source every second. Activity is measured in becquerels (Bq).

The half-life of a radioactive source is the time taken for its activity to halve.

Alpha, beta and gamma radiation

Types of radiation

An atom can be thought of as a small nucleus surrounded by orbiting electrons. The nucleus of the atom contains particles called protons and neutrons.

Complete the labelling of the diagram to show the nucleus, protons and orbiting electrons.

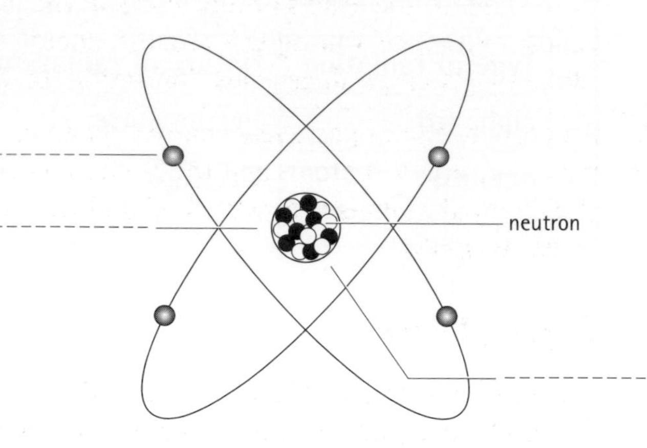

REMEMBER The three types of radiation transfer energy from the nucleus when they are emitted. The energy transferred by the nucleus enables it to become more stable.

The nuclei of the atoms in some elements are unstable. The nuclei become stable by emitting radiation. The radiation emitted by the nuclei of unstable atoms can be of three types. The three types are called alpha radiation (α), beta radiation (β) and gamma radiation (γ).

The alpha radiation is slow-moving helium nuclei. They travel at about 5% of the speed of light.

Beta radiation is fast-moving electrons. They can travel at about 90% of the speed of light.

Gamma radiation is high-energy electromagnetic waves. The electromagnetic waves are emitted in short bursts by a radioactive atom. These short bursts of gamma radiation travel at the speed of light.

Absorption

The energy of the three radiations is absorbed when the radiations pass through material. The amount of energy that is absorbed depends on the type of radiation and the type of material.

The range of the alpha radiation in an absorbing material is less than that of beta or gamma. The alpha radiation transfers more energy to the absorber than beta or gamma radiation. Alpha radiation can be absorbed by paper or by the thickness of the skin or by a few centimetres of air.

Beta radiation is more penetrating than alpha radiation. It can pass through the skin, but it is absorbed by a few centimetres of body tissue or a few millimetres of aluminium.

Gamma radiation is the most penetrating of the three radiations. It can easily penetrate body tissue. It requires a few centimetres of lead or about 1 metre of concrete to absorb it.

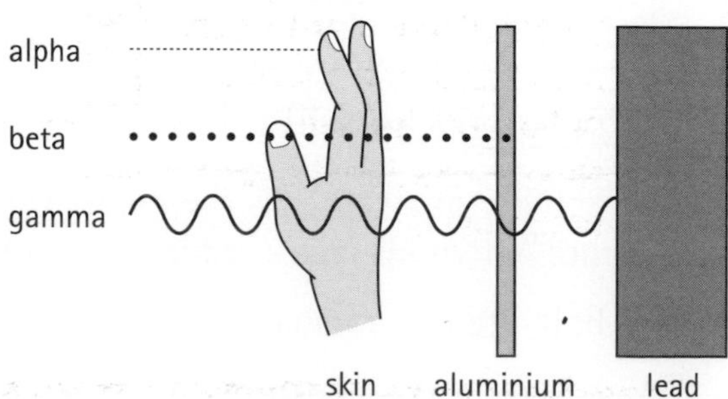

Ionisation

When radiation passes through the absorbing material, electrons are stripped from some of the atoms or molecules in the absorbing material. The process is called ionisation. During this process, energy is transferred to the absorber. Alpha radiation produces most ionisation, beta next and gamma least.

An atom becomes positively charged when an electron is removed from it. The positively charged atoms and the negatively charged electrons are called 'ions'. The charged atom is a 'positive ion' and the negatively charged electron is a 'negative ion'.

Detection

The ionising property of alpha, beta and gamma radiation is used in the Geiger-Müller (GM) tube as a means of detecting the radiation.

The GM tube is a hollow cylinder filled with a gas at low pressure. The tube has a thin window made of mica at one end. There is a central electrode inside the GM tube. A voltage supply is connected across the casing of the tube and the central electrode as shown in the diagram.

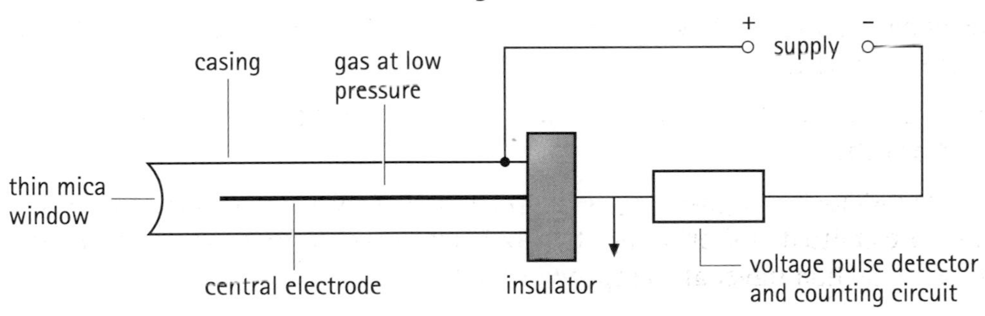

When alpha, beta or gamma radiation enters the tube, it produces ions in the gas. The ions created in the gas enable the tube to conduct. A current is produced in the tube for a short time. The current produces a voltage pulse. Each voltage pulse corresponds to one ionising radiation entering the GM tube. The voltage pulses are counted and recorded.

Practice questions

1) Suggest why the window of a GM tube is made of thin material.
2) The diagram shows how far two radiations travel when different absorbers are placed in front of a radioactive source. Name radiation X and radiation Y.

X
Y
paper
2 mm aluminium
2 cm lead

Activity and half-life

Activity

The activity of a radioactive source is the number of radiations it emits every second. Activity is measured in becquerels (Bq). An activity of 1 Bq means that every second either an alpha particle, a beta particle or a gamma ray is emitted. The activity of a radioactive source is sometimes measured in kilobecquerels (kBq) or megabecquerels (MBq).

The activity of a radioactive source decreases with time. If you plot a graph of how the activity of a radioactive source decreases with time, you get a smooth curve.

All radioactive sources give graphs of a similar shape to this one.

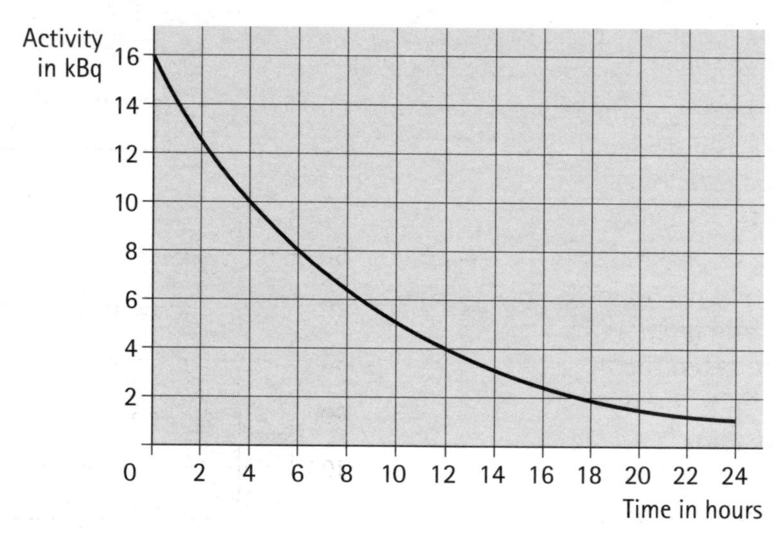

Take readings from the graph to complete the following table.

Activity decrease from	Time taken for decrease (hours)
16 kBq to 8 kBq	6
10 kBq to 5 kBq	6
8 kBq to 4 kBq	
6 kBq to 3 kBq	
4 kBq to 2 kBq	

You can see from the graph that the time taken for the activity to reduce by a half is constant. Knowing the value of this constant, means that you can predict what the activity of the source will be at a particular time.

Half-life

The time taken for the activity of a radioactive source to decrease by half is called the 'half-life' of the source.

The half-life of the source in the graph on page 86 is 6 hours. So after 6 hours (one half-life) the activity is $\frac{1}{2}$ of what it was originally. After 12 hours (two half-lives) it is $\frac{1}{4}$ of the original value. After 18 hours (3 half-lives) it is $\frac{1}{8}$ of the original activity.

> **REMEMBER** At Credit Level, you need to know how to measure the half-life of a radioactive source.

After 10 half-lives, approximately what fraction of the original activity of a radioactive source will remain?

A $\frac{1}{2}$ *B* $\frac{1}{10}$ *C* $\frac{1}{20}$ *D* $\frac{1}{100}$ *E* $\frac{1}{1000}$

The half-lives of radioactive elements can range from a fraction of a second to thousands of years.

The half-life of a radioactive source could be measured as follows.

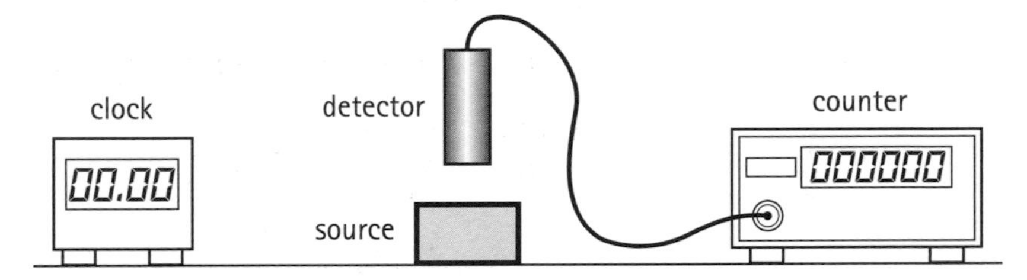

Firstly, you measure the background count rate using a GM tube connected to a counter. The count rate from the source is measured at regular fixed intervals over a period of time. You then subtract the background count rate from each measurement to get the actual count rate of the source. A graph of the count rate of the source against time is plotted. From the graph, you can measure the time taken for the count rate to fall by half. You need to make a number of measurements to calculate the average value. The average value is the half-life of the radioactive source.

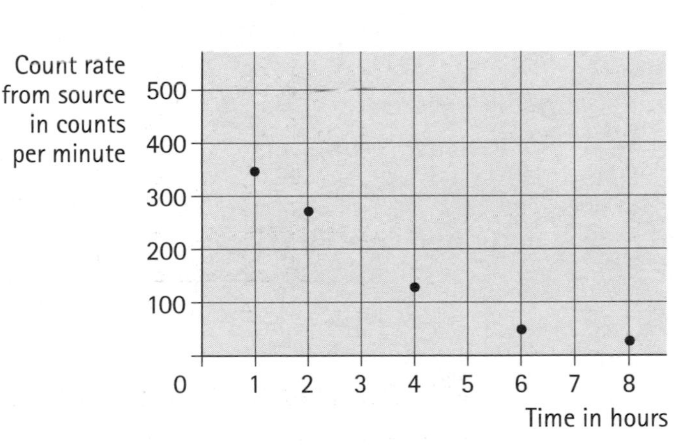

Practice questions

1) These data give information on the count rate measured at different times over a period of 8 hours. The background count rate is 20 counts per minute.

Time (hours)	0	1	2	3	4	5	6	7	8
Counts per minute	520	360	272	215	145	110	82	63	52
Count rate of source		340	252		125		62		32

a) Complete the table. Plot the graph of the count rate of the source against time.

b) Use your graph to calculate the half-life of the radioactive material.

2) A radioactive beta source emits 5000 beta particles in 4 seconds. What is the activity of the source?

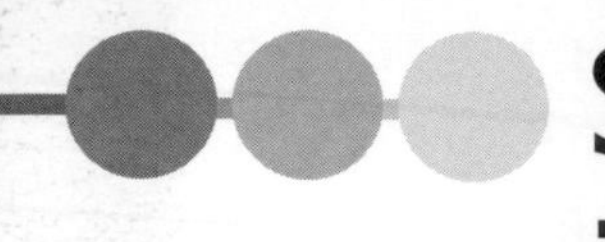

Safety and uses of radiation in medicine

Safety

When working with radioactive sources, people use safety precautions to keep the radiation they are receiving to a minimum.

REMEMBER Make sure you know about the safety precautions that are needed when working with radioactive sources.

Symbol for radioactive source

One way of limiting the dose from a radioactive source is to keep the source as far away from the user as possible. This can be done by using tongs to handle the source or by wearing special gloves.

Radioactive sources should be kept away from the body and never brought close to the eyes.

Another way is to use shielding. Sources can be stored in lead-lined containers. The shielding acts as an absorber of the radiation and so reduces the amount of radiation reaching anyone close to the source.

REMEMBER At Credit Level, you should know that the radiation dose received by a person depends on: the type of radiation, the length of time exposed to the radiation and the type of tissue affected by the radiation.

It is usual to limit the amount of time that a person is close to the source of radiation. The radiation received by a person working with radioactive sources (e.g. a hospital worker) can be monitored using film badges. The film in the badge is sensitive to radiation. The amount of blackening that develops on the film is a measure of the dose received by the person wearing the badge. Film badges are checked regularly to ensure that the dose received does not exceed safety limits.

The radiation dose received by body tissue depends on the type of radiation absorbed, i.e. whether the radiation is alpha, beta, gamma or some other type of radiation such as X-rays. The radiation dose received also depends on the energy of the radiation.

A way of expressing the radiation dose received from different sources is in terms of a quantity called 'dose equivalent'. Dose equivalent is measured in sieverts (Sv). A dose of 1 Sv from an alpha radiation source is equivalent to a dose of 1 Sv from a beta radiation source or any other source of radiation.

The risk of damage to body tissue from absorbed radiation depends on the number of sieverts the person receives. The risk of damage also depends on the type of body tissue absorbing the radiation – some body tissues are more susceptible to harm from radiation than others.

Uses of radiation in medicine

Gamma radiation can be used to sterilise medical equipment such as syringes, bandages and scalpels. The gamma radiation kills harmful bacteria that may have contaminated the equipment.

Radioactive material can also be used in medical diagnosis. Gamma radiation is a useful tool for finding out if parts of the body are working as they should. Radioactive material, such as technetium 99 or iodine 123, can be made into a solution and injected into the body.

The radioactive solution is prepared under sterile conditions by medical technicians. The technicians ensure that the activity of the solution is at a level which is safe for injection into the body. Once inside the body, the blood carries the radioactive solution to the area which is to be examined.

The gamma radiation passes through the body tissue and is detected by a special device called a gamma camera. The gamma camera is an array of a large number of gamma detectors. Each detector is linked to a computer. Each detector in the array enables the computer to form an image of the part of the body that is being examined.

This diagram shows the lungs as viewed by a gamma camera. The black area on the right lung shows that the blood flow to this part of the lung is restricted.

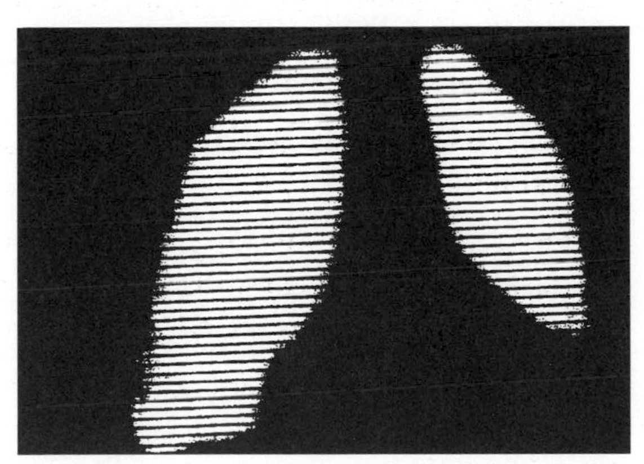

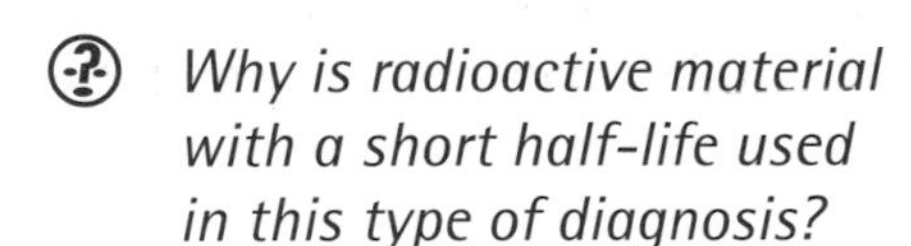

Why is radioactive material with a short half-life used in this type of diagnosis?

Practice questions

1) Alpha-emitting radioactive material is never injected into the body in order to produce images of body parts.

   Give two reasons why alpha emitters are not suitable.

2) When hospital technicians are making up samples of radioactive solutions for use with patients, they wear detector film badges on the fingers of their gloves as well as on their laboratory clothing.

   Give a reason for this.

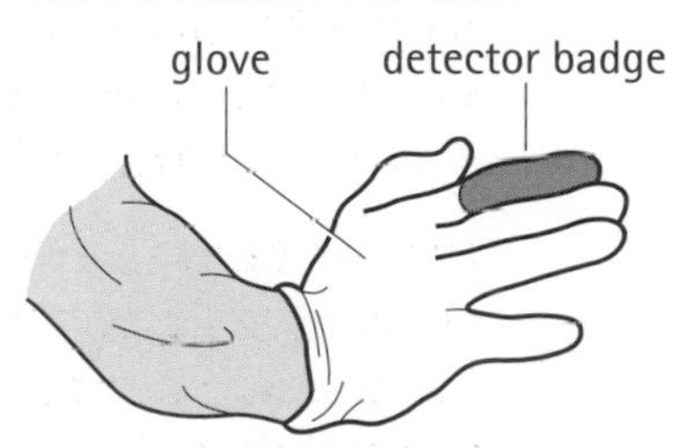

Examination questions

Try these two questions from past exam papers. Question 1 is from a **General Level** paper and question 2 is from a **Credit Level** paper. Spend about 5 minutes on question 1 and no more than 10 minutes on question 2.

When you have finished, turn to page 96 for the answers and the marking guide.

1) Radioactive sources are often stored in lead-lined containers as shown.

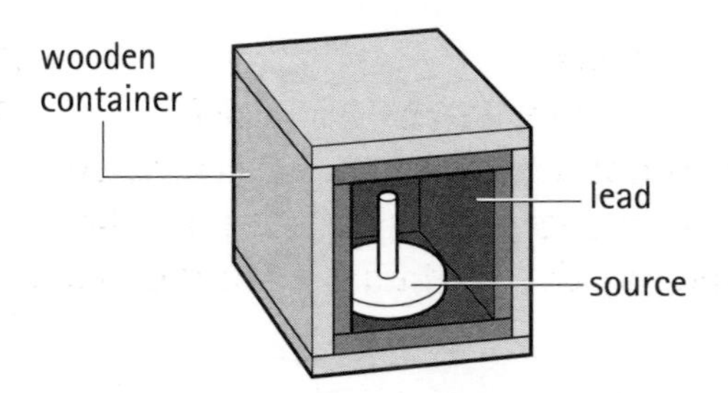

a) Explain why the container is lead-lined. **(1)**

b) Describe **two** safety precautions which should be taken when using the source. **(2)**

c) The source is placed in front of a Geiger-Müller tube and counter as shown below. The counter registers 600 counts in one minute.
An identical experiment is repeated two years later using the same source. The reading on the counter is now 500 counts in one minute.
Explain why the count rate has decreased. **(1)**

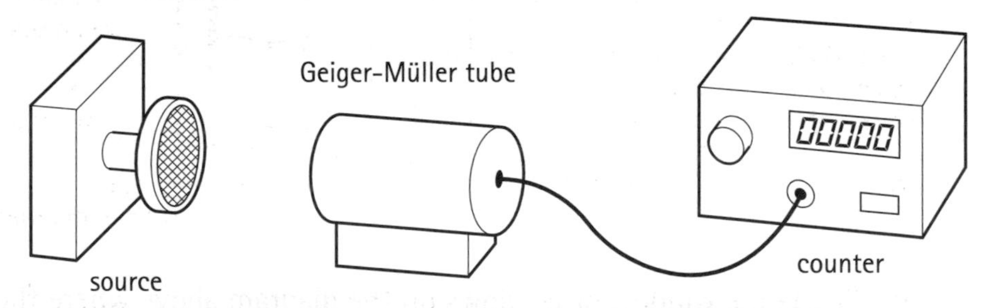

2) a) A hospital uses radioactive technetium in the diagnosis of tumours. The technetium is injected into the patient. The label on a sample which is delivered to the hospital is shown below.

TECHNETIUM	
Date of delivery	15/5/1997
Time of delivery	1.00pm
Half-life	6 hours
Activity on delivery	600 MBq
Type of radiation	Gamma

i) What is meant by the term 'half-life'? **(1)**

ii) Why is a sample with a short half-life used in diagnosis? **(1)**

iii) If the sample of technetium is not used, the hospital is allowed to dispose of it. This is permitted once its activity has fallen below 75 MBq. Show, by calculation, the date and time when the sample will be ready for disposal. **(3)**

b) The effect of radiation absorbed by living materials depends on a number of factors. Name **two** of the factors. **(2)**

c) Members of the hospital staff wear film badges to monitor any radiation to which they may be exposed. The film is contained in a plastic holder with windows of different materials as shown in the diagram. The whole badge is protected from light.

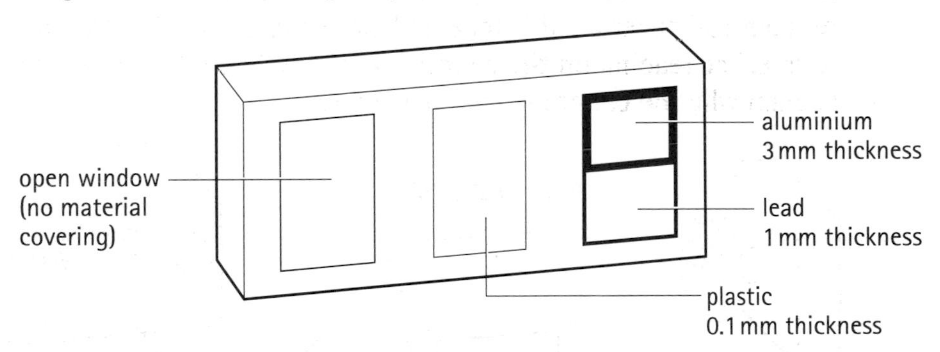

i) Shade the window or windows on the diagram above where the film will be affected if the wearer is accidentally exposed to the radiation from the Technetium source. **(1)**

ii) Describe how the badge is used to indicate how much radiation has been received. **(1)**

Answers to ? questions

Forces and motion

Page 10 Acceleration is important at the start of the race. A car with a good acceleration can get clear of the other cars and take up a good position at the first bend. Top speed is important on long straights. The car with a high top speed can put more distance between itself and the other cars.

Page 11 We cannot say whether the car broke the speed limit. It may have been travelling at a fairly steady speed throughout the journey, in which case the speed limit would not have been broken. Alternatively, it may have been travelling very slowly at some point during its journey. Therefore, it must have travelled very quickly at other times in order to maintain an average speed of 30 m/s (66 miles per hour). If this were the case, it is likely that the speed limit was exceeded, but we cannot tell.

Page 14 The left-hand arrow.

Page 16 The A force: the balloon pushing on the air in the balloon. The B force: the air in the balloon pushing on the balloon.

Page 18
Gravitational field strength on the Moon

$= \frac{1}{6}$ gravitational field strength on Earth

$= \frac{1}{6} \times 10$

$= 1.6$ N/kg

Page 19 The weight of the feather is much less than that of the coin. The size of the air resistance on the feather is likely to be greater than the air resistance force on the coin. So the unbalanced force on the feather is less than the unbalanced force on the coin. Therefore, the downward acceleration of the feather is less than that of the coin.

Page 21 It would move away from you, but would stay level with you.

Energy

Page 27 450 light bulbs and 45 electric fires

Page 30 At Q energy transfer is E_p to E_k; at S energy transfer is E_k to E_p

Page 34 Turning up the gas does not increase the temperature of the water when it is already at its boiling point. The temperature of the water surrounding the egg remains at 100°C, so the egg does not cook more quickly.

Page 35 Vaporisation of water – about 7 times more energy.

Electricity

Page 40 Ski lift = battery; skiers = charges; path followed by skiers = lamp; ski lift and path followed by skiers = circuit; current = number of skiers passing per second

Page 42 2A and 2V

Page 43 2 branches; current from battery = 3A; V_2 = 6V; battery voltage = 6V

Page 45 The voltage and current are directly proportional, so $\frac{V}{I}$ is a constant for all values of V and I.

Page 47 The ring circuit is used for more powerful appliances. So it normally draws a larger current from the supply. Therefore, the fuse needed for the lighting circuit has a lower rating

Electronics

Page 58 a) microphone; b) thermistor; c) solar cell; d) LDR

Page 60 As a parking light, a security light or a street light.

Page 61 An OR gate is so called because it opens when one input **or** the other is high. A NOT gate is so called because when its input is high its output is **not** high (i.e. low).

Page 63 0 to 9

Waves

Page 77 When the light is incident at right angles on the boundary between the two media.

Radioactivity

Page 87 After 10 half-lives, radioactive material is approximately one-thousandth as active as at the start (it has been divided in half 10 times).

Page 89 Activity falls quickly to a low level and therefore reduces the radiation dose received.

Answers to practice questions

Forces and motion

Page 11

a) 0 m/s
b) 26 m/s
c) 3.25 m/s²
d) 32.5 m

Page 13

1 a) 22.5 m/s
b) acceleration = 1.5 m/s²; deceleration = 2.25 m/s²
c) 506.25 m
2 a) B b) C c) D
d) E e) A

Page 17

1 2 m/s²
2 a) 2 N (acting to the left)
b) 0.5 m/s²
3 a) P and R
b) 0.2 m/s²
c) 1500 N
d) 2 m/s
e) Q and S
4 balanced; unbalanced; deceleration; speed; force

Page 19

1 a) 7500 N b) 7500 N
2 a) 600 N
b) 50 N downwards
c) 0.83 m/s²

Page 21

1 a) Path of ball seen by you:

b) Path of ball seen by observer:

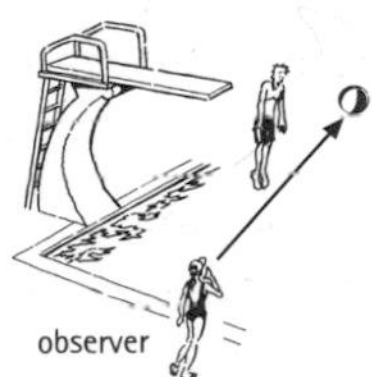

2 a) 2.5 N b) 10 m/s²
c) 14 m

d)

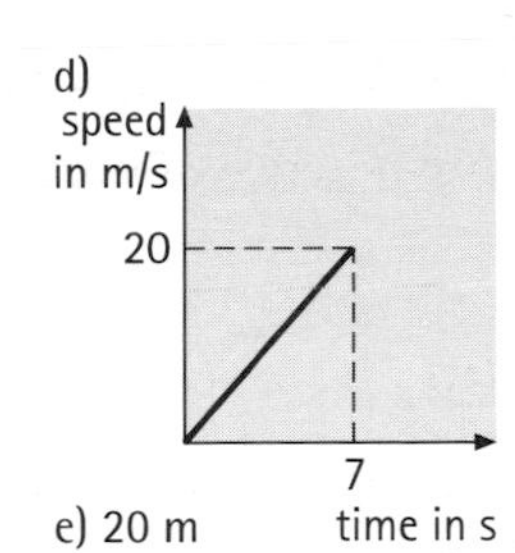

e) 20 m

Energy

Page 27

1 a) 400 N b) 3200 J
2 a) 3.3 m/s b) 90 kJ
c) 60 N
3 a) 75 kJ b) 75 kJ
c) 250 W

Page 29

1 a) 500 N b) 2500 J
c) 2500 J
2 a) 120 kJ b) 6000 W
3 Cricket ball. Kinetic energy depends on speed and mass. Speed is same for both balls. Mass of cricket ball is greater than that of tennis ball.
4 10 m/s

Page 31

1 a) 5 m b) 2500 J
c) 10 m/s
2 a) 1920 J b) 8 m/s

Page 33

1 42.5 kJ
2 a) 266 475 J
b) 177.65 s

Page 35

1 11300 J
2 2672 J

Electricity

Page 41

a)

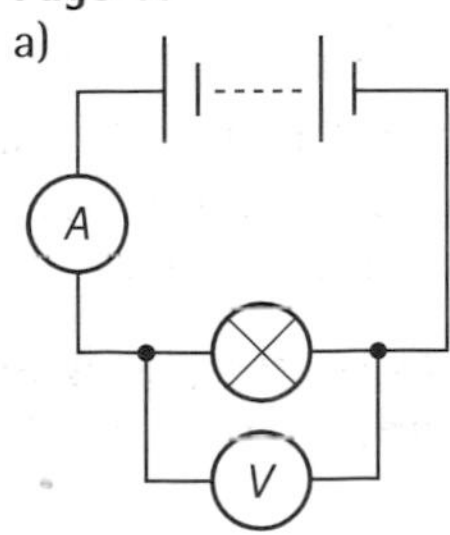

b) 3 V c) 6 C

Page 43

1

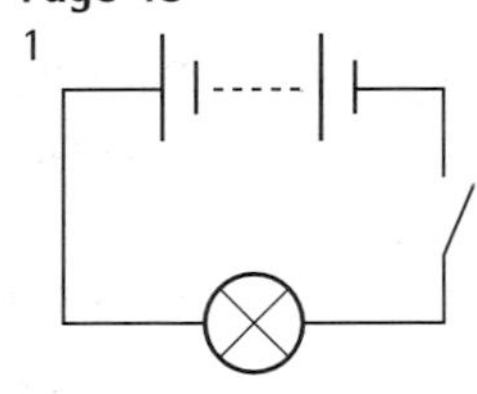

2

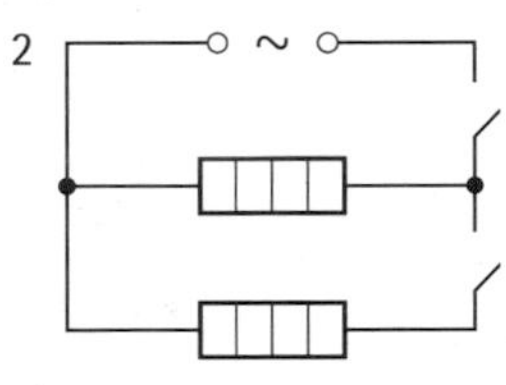

3 a) parallel
b) i) rear and side;
ii) none are lit;
iii) all are lit
c) 10 A

Page 45

1 3 volts
2 a) B and D (parallel); A and C (series)
b) R_A = 10 ohms; R_B = 2.5 ohms

Page 47

a) 6.52 A b) 13 A
c) safety; prevents body of kettle becoming live if a fault develops.

Page 49

a) Rotation of the bicycle wheel causes magnet to rotate. Rotating magnet produces changing magnetic field near the coil. This induces a voltage in the coil.
b) Alternating current.
c)

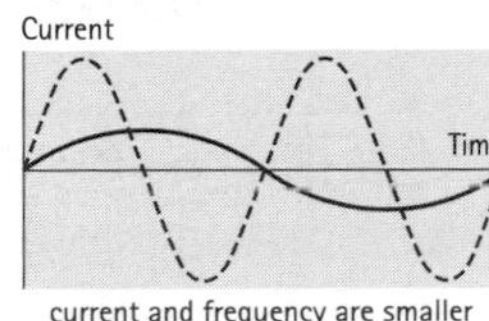

current and frequency are smaller

Page 51

1 23 V
2 a) 12 A
b) i) 25 W ii) 12.5 A
3 Car battery supplies d.c.; transformer requires a.c.

Electronics

Page 57

1 input: keyboard or mouse; output: VDU
2 amplifier

Page 59

1 133.3 ohms
2 Pressing switch discharges capacitor, voltage across resistor increases from zero to the voltage of the supply. On releasing switch, capacitor starts to charge, current in resistor decreases, voltage across resistor decreases to zero.

Page 61

Circuit 1: NOT;
Circuit 2: OR;
Circuit 3: AND

Waves

Page 69

1 a) 4 Hz b) 5 m/s
c) 1.25 m
2 a) 0.5 m b) 2 m
c) 6 m/s

Page 71

1 0.25 m
2 1500 m/s

Page 73

1500 m

Page 75

1 LW: long wave; MW: medium wave; SW: short wave; VHF: very high frequency; UHF: ultra high frequency,
2 Speed of light from flash is about 1 million times greater than the speed of sound from the thunder.
3 X and Y

Page 79

1 Brighter image because more light is admitted into the telescope tube.
2 Reduces the size (smaller magnification).

Radioactivity

Page 85

1 The thin window allows alpha radiation into the tube and enables it to be detected.
2 X is beta radiation; Y is gamma radiation.

Page 87

1 a) Count rate of source at 0 h (500) 3 h (195) 5 h (90) 7h (43)
b) Half-life is 2 hours.
2 1250 Bq

Page 89

1 Alpha emitter inside the body is dangerous and harmful to body cells. Also alpha radiation would not penetrate the body tissue to enable detection on the outside of the body.
2 The badge on the glove is close to the source of the radiation and so will provide a better indication of the dose actually received by the hand.

Answers to examination questions

Marking guide

The marks shown in brackets are the marks you award yourself for each step in the answer, or key word or phrase, that you get correct.

In calculations, if you gave a wrong answer and did not show your working, award yourself zero marks for the question. If you used an incorrect formula, award yourself zero marks.

Marks are awarded for partial success in the examination. You might see part of the way to solving a problem. Show this to the examiner.

Make sure you show your working.

Forces and motion, pages 22–23

1) a) acceleration = change in speed/time taken [half mark for correct formula]

$= \frac{60}{12}$ [half mark for correct substitution]

= 5 miles per hour per second [1 mark, deduct half mark if unit is wrong/missing]

b) i) Car C [half mark] because the final speed of the car C in the trial run [half mark] is greater than [half mark] the top speed of car A or car B. [half mark]

ii) accelerating OR speed is increasing [1 mark]

c) It would increase the time. [1 mark]

2) a) i) 8 seconds [1 mark, deduct half mark if unit is wrong/missing]

ii) area under graph = distance travelled [1 mark for correct formula]

$= (35 \times 8)$ [half mark] $+ (\frac{1}{2} \times 7 \times 35)$ [half mark] = 402.5 m [1 mark, deduct half mark if unit is wrong/missing]

iii) deceleration = change in speed/time taken [half mark for correct formula]

$= \frac{35}{7}$ [half mark for correct substitution] = 5 m/s² [1 mark, deduct half mark if unit is wrong/missing]

iv) unbalanced force = mass × acceleration (or deceleration) [half mark for correct formula]

$= 380 \times 5$ [half mark for correct substitution] = 1940 N [1 mark, deduct half mark if unit is wrong/missing]

[Award full marks if the value you gave was wrong but is consistent with the value you calculated in iii)]

b) i) During BC, the mass m being pushed increases. [1 mark]
The pushing force F is decreased. [1 mark]

ii) Line with slope (gradient) less than that of BC [1 mark]

Energy, pages 36–37

1) a) $E = cm\Delta T$ [half mark for correct formula]

$= 1000 \times 80 \times 20$ [half mark for correct substitution of data]

= 1600 kJ [1 mark, deduct half mark if unit is wrong/missing]

b) The greater the temperature difference, the greater the rate of heat transfer. [1 mark] The temperature difference between inside and outside is greater than the temperature difference between rooms. [1 mark]

c) i) X [1 mark] because the temperature stays higher for longer. [1 mark];

ii) 5 °C [1 mark, deduct half mark if unit is wrong/missing]

2) a) $E_p = mgh$ [half mark for correct formula]

$= 50 \times 10 \times 11.25$ [half mark for correct substitution of data]

= 5625 J [1 mark, deduct half mark if unit is wrong/missing]

b) $E_p = \frac{1}{2} mv^2$ [half mark for correct formula]

$5625 = \frac{1}{2} \times 50 \times v^2$ [half mark for correct substitution of data]

$v^2 = 225$ [half mark]

$v = 15$ m/s [half mark for taking square root]

[Award full marks if the value you gave was wrong but is consistent with the value you calculated for E_p in a)]

c) i) area under graph = distance travelled [half mark for correct formula]
$= \frac{1}{2} \times 7 \times 3$ [half mark] + 13×3 [half mark] + $\frac{1}{2} \times 13 \times 9$ [half mark]
= 10.5 + 39 + 58.5 = 108 m [1 mark , deduct half mark if unit is wrong/missing]
ii) heat produced = work done against friction
$E_h = F \times d$ [half mark for correct formula]
$2025 = F \times 108$ [half mark for correct substitution of data]
$F = 18.75$ N [1 mark, deduct half mark if unit is wrong/missing]
[Award full marks if the value you gave is wrong but it is consistent with the answer you gave for part c)i)]

Electricity, pages 52–53

1) a) $P = \frac{V}{I}$ [half mark for correct formula]

$= 12 \times 10$ [half mark for correct substitution of data]

= 120 W [1 mark, deduct half mark if unit is wrong/missing]

b) $R = \frac{V}{I}$ [half mark for correct formula]

$= \frac{12}{10}$ [half mark for correct substitution of data]

= 1.2 Ω [1 mark, deduct half mark if unit is wrong/missing]

c) less than [1 mark]

2) a) i) nuclear fuel OR hydroelectricity [1 mark];
ii) solar OR wind OR waves OR geothermal [1 mark];
iii) no pollution OR environmentally friendly OR it is renewable [1 mark]

b) i) $\frac{V_s}{V_p} = \frac{n_s}{n_p}$ [half mark for correct formula]

$\frac{400\,000}{25\,000} = \frac{n_s}{20\,000}$ [half mark for correct substitution of data]

$n_s = 320\,000$ [1 mark, deduct half mark if unit is given]

ii) High voltage reduces power losses OR energy losses [1 mark]

3) a) i) The rotating magnet produces a changing magnetic field in the vicinity of the coil. [half mark] The changing magnetic field induces a voltage in the coil. [half mark];
ii) Use stronger magnet. [1 mark] Increase the number of turns in the coil. [1 mark]

b) i) Part X is a transformer. [half mark] It steps up the voltage from the generator. [half mark]; Part Y is a transformer. [half mark] It steps down the voltage to a level suitable for use by the consumer. [half mark];
ii) resistance of power line = 2 × 100 = 200 Ω [1 mark]; power loss in line = $I^2 R$ [half mark for correct formula] = $(200)^2 \times 200$ [half mark for correct substitution of data] = 8 000 000 W [1 mark, deduct half mark if unit is wrong/missing]
[You can earn 2 partial marks if you calculated the resistance of the power line incorrectly, but you used this value correctly and consistently in calculating the power loss.]

Electronics, pages 64–65

1) a) i) output device is a loudspeaker [half mark] input device is a microphone [half mark];
ii) electrical energy to sound energy [1 mark]

b) the output voltage is 50 000 times the input voltage OR $\frac{\text{output voltage}}{\text{input voltage}} = 50\,000$ [1 mark]

c) 1000 Hz [1 mark, deduct half mark if unit is wrong/missing]

d) energy = power × time [half mark for correct formula] = 120 × (5 × 60) [half mark for correct substitution of data] = 36 000 J [1 mark, deduct half mark if unit is wrong/missing]

2) a) i) AND [1 mark]

ii) When light shines, the logic output of the sensor = 0 [half mark]. When light shines on sensor, one of the inputs to the AND gate is 0 [half mark]. So in this condition output of AND gate = 0 [half mark] Therefore no count is recorded. [half mark]
iii) 5 [1 mark]

b) i) Time = 8 × 0.01 = 0.08 s [1 mark, deduct half mark if unit is wrong/missing]
ii) The length of the car that interrupts the light beam [1 mark]

iii) He should use speed = $\frac{\text{length of car}}{\text{time recorded}}$ [1 mark]

(Award only half mark if your answer was: speed = $\frac{\text{distance}}{\text{time}}$)

c) i) Increasing the size of the capacitance lowers the frequency of the oscillation. [1 mark];
ii) less accurate [1 mark] [If your answer to c) part i) was decreases the frequency, award 1 mark for 'more accurate'.]

Waves, pages 80–81

1) a) frequency = number of waves produced per second [half mark]; 24 waves [half mark] are produced in 60 seconds [half mark] Therefore 0.4 of a wave is produced in 1 second [half mark] Therefore the frequency of waves is 0.4 hertz

b) 0.4 × 4 [1 mark]
= 1.6 m/s [1 mark, deduct half mark if unit is wrong/missing]

c) amplitude = 0.25 m [1 mark, deduct half mark if unit is wrong/missing]

2) a) Refraction is the change in speed OR change in direction [half mark] as light passes from one medium to another. [half mark]

b) Use a lens with a large curvature OR use lens of shorter focal length. [1 mark] Lens 1 represents the cornea [half mark] and so it provides most of the refraction. [half mark]

c) i) Q circled [1 mark]
ii) concave shape [1 mark] [Award it if you drew a concave shaped lens.]
iii) power of lens = $\frac{1}{\text{focal length}}$ [half mark for correct formula]

$4 = \frac{1}{f}$ [half mark for correct substitution of data]

$f = 0.25$ m [1 mark, deduct half mark if unit is wrong/missing]

iv) reduces the focal length [1 mark]

Radioactivity, pages 90–91

1) a) To absorb the radiation that is emitted by the source. [1 mark]

b) Handle with tongs OR do not point towards the body OR minimise the time spent near the source [2 marks: any two; 1 mark each]

c) The activity of a radioactive source decreases with time. [1 mark]

2) a) i) Half-life is the time taken for the activity of the source to reduce by half OR time taken for half of the original number of radioactive atoms to decay. [1 mark];
ii) A short half-life means that the activity of the source decreases quickly. [half mark] This means that any risk to the patient, resulting from exposure to a radioactive source, is reduced. [half mark];
iii) Activity reaches 75 MBq after 3 half-lives. [1 mark] Therefore time that elapses is 3 × 6 = 18 hours
[1 mark]; So disposal time = 7 am on 16/5/97 [1 mark]

b) Type of radiation OR type of absorbing tissue OR energy of radiation [2 marks: any two; 1 mark for each]

c) i) All of the windows would be shaded on the diagram. [1 mark];
ii) The amount of blackening of the film behind each window is an indication of the duration of exposure to the radiation. [1 mark]